# Kindle Fire Survival Guide

# from MobileReference

## By Toly K

## Table of Contents

# Getting Started

## Table of Contents

## 1. Button Layout

The Kindle Fire has one button, a headphone jack, and a micro-USB port. The touchscreen is used to control all functions on the Kindle Fire, with the exception of turning on the device. The Kindle Fire has the following buttons:

*Figure 1: Front View*

**Touchscreen** - Used to control all functions on the Kindle Fire.

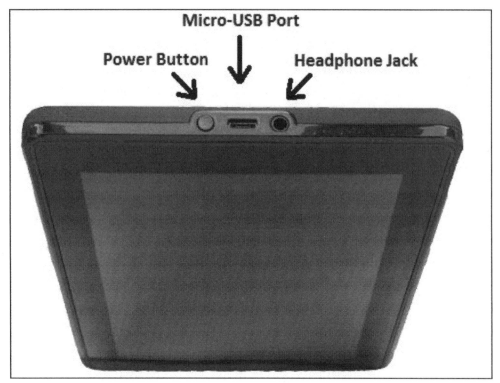

*Figure 2: Bottom View*

**Headphone Jack** - Allows headphones to be connected.

**Micro-USB Port** – Allows the Kindle Fire to be connected to a computer to transfer data.

## 2. Charging the Kindle Fire

For optimal battery life, charge the Kindle Fire fully before using it for the first time. Using the included charging adapter, plug in the tablet. The **Power** button at the bottom of the device illuminates an orange color. When the Kindle Fire is done charging, the Power button turns green.

*Note: The Kindle Fire cannot charge while it is turned off.*

# 3. Turning the Kindle Fire On and Off

To turn the Kindle Fire **ON**, press the **Power** button at the bottom of the device and immediately release it. The Power button illuminates a green color and the Kindle Fire turns on. When the device is fully turned on, the Lock screen appears.

To turn the Kindle Fire **OFF**, press and hold the **Power** button for two seconds. The Shut Down confirmation appears, as shown in **Figure 3**. Touch the Shut Down button. The Kindle Fire turns off.

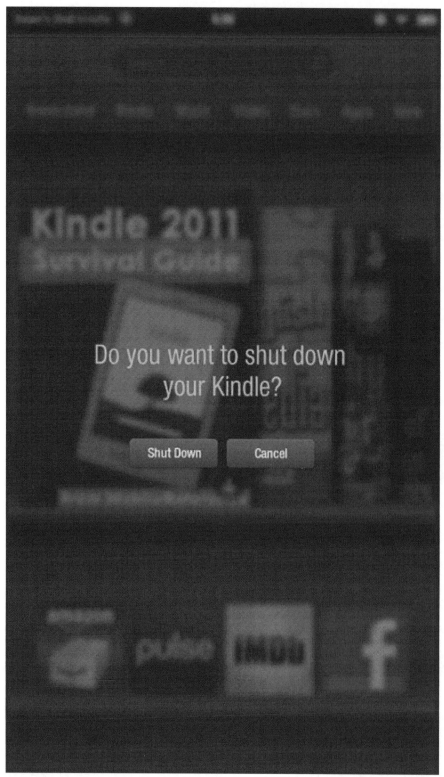

*Figure 3: Shut Down Confirmation*

# 4. First-Time Setup

The first time the Kindle Fire is turned on, it must be set up. To perform first-time setup:

*Note: If your Kindle fire came registered to the account you used to purchase it, steps 6 and 7 below are not required.*

1. Press the **Power** button at the bottom of the screen and immediately release it. The Kindle Fire turns on and the Welcome screen appears, as shown in **Figure 4**.
2. Touch a Wi-Fi network in the list. The Password prompt appears, as shown in **Figure 5**.
3. Type the password for the Wi-Fi network you selected. The password is usually found on the back or side of your wireless router.
4. Touch the [Connect] button. The Kindle fire connects to the Wi-Fi network and the Time Zone screen appears, as shown in **Figure 6**.
5. Touch the time zone that applies to your location and then touch the [Continue] button. The Registration screen appears, as shown in **Figure 7**. If your Kindle Fire came registered, skip steps 6 and 7.
6. Type the email address associated with your Amazon account and then touch the [Next] button. The Password field is highlighted.
7. Type the password associated with your Amazon account and then touch the [Register] button. The Kindle Fire is registered and a confirmation appears.

*Note: The device may need to download additional software before you can use it.*

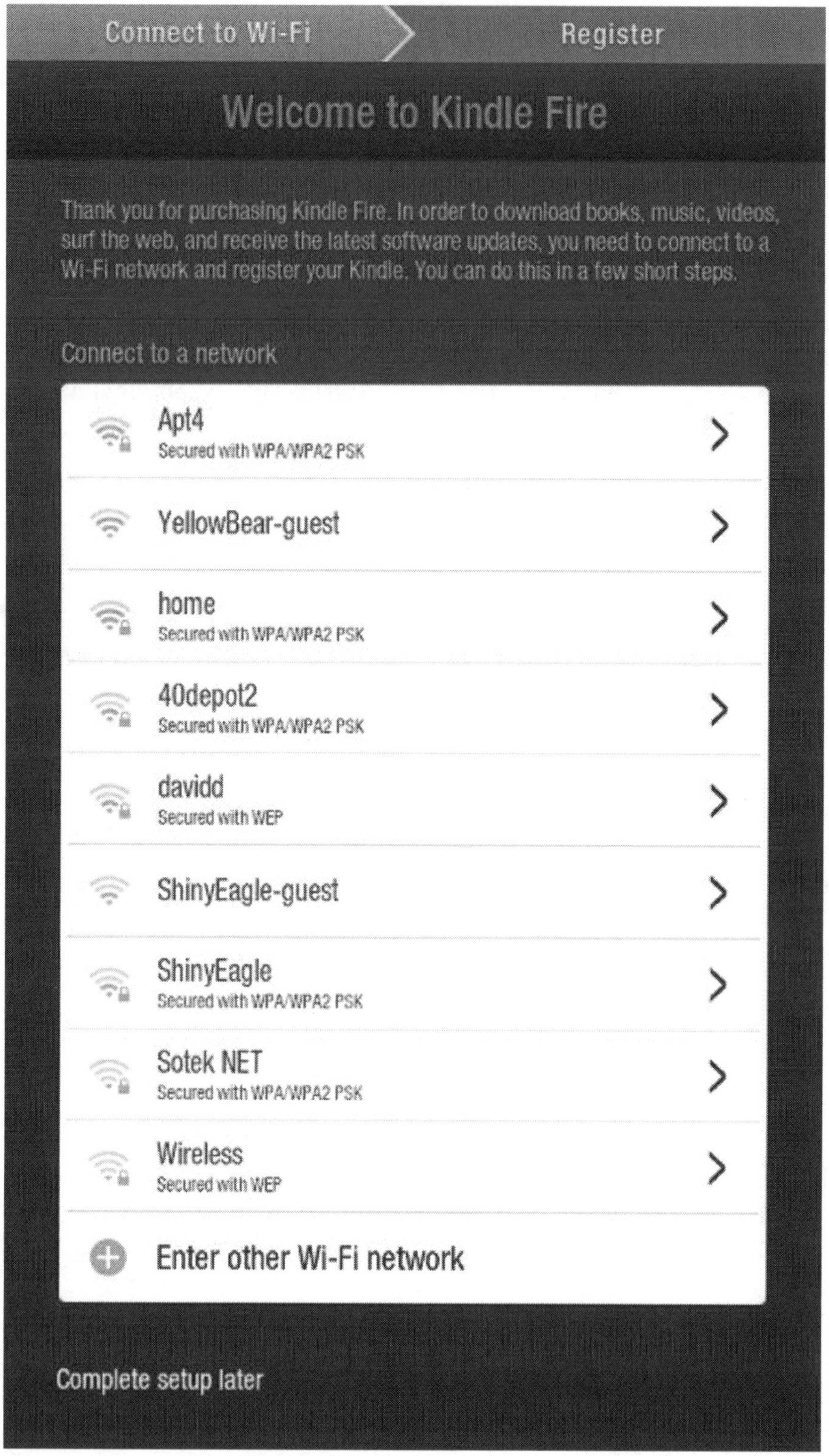

Connect to Wi-Fi      Register

# Welcome to Kindle Fire

Thank you for purchasing Kindle Fire. In order to download books, music, videos, surf the web, and receive the latest software updates, you need to connect to a Wi-Fi network and register your Kindle. You can do this in a few short steps.

Connect to a network

Apt4
Secured with WPA/WPA2 PSK                >

YellowBear-guest                          >

home
Secured with WPA/WPA2 PSK                >

40depot2
Secured with WPA/WPA2 PSK                >

davidd
Secured with WEP                         >

ShinyEagle-guest                          >

ShinyEagle
Secured with WPA/WPA2 PSK                >

Sotek NET
Secured with WPA/WPA2 PSK                >

Wireless
Secured with WEP                         >

⊕  Enter other Wi-Fi network

Complete setup later

*Figure 4: Welcome Screen*

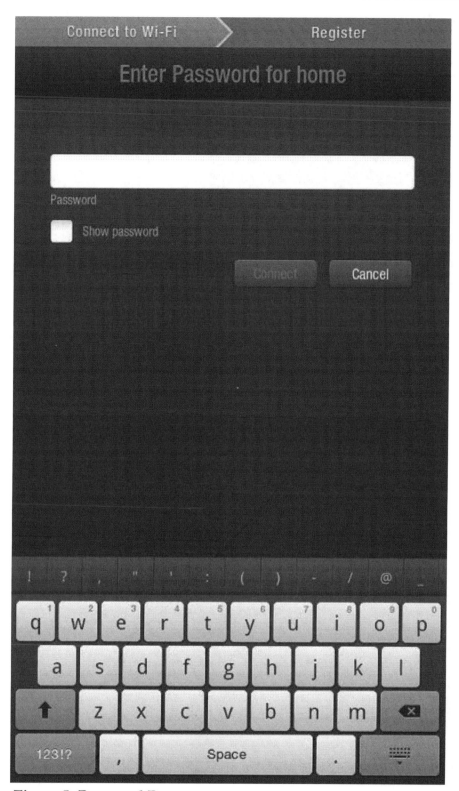

*Figure 5: Password Prompt*

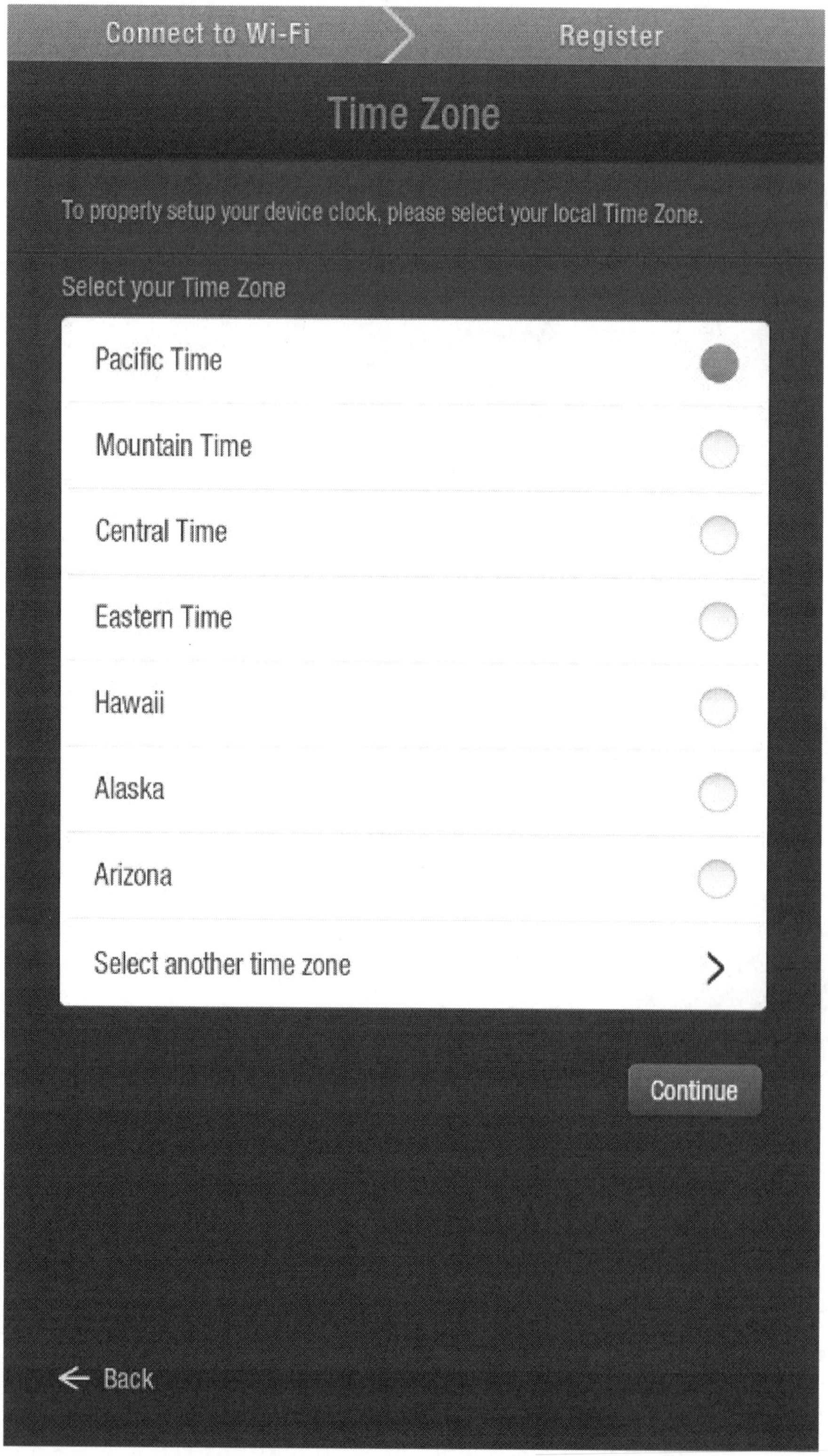

*Figure 6: Time Zone Screen*

*Figure 7: Registration Screen*

# 5. Navigating the Screens

There are many ways to navigate the Kindle Fire. These are just two of the methods:

- Touch the button at the bottom left of the screen to return to the Library at any time. Any application or eBook will be in the same state when it is re-opened.
- Touch the button to return to the previous screen or menu, or to hide the keyboard.

# 6. Managing Favorites in the Library

In order to access the items you use most often on the Kindle Fire, add them to your Favorites. The Favorites are displayed at the bottom of the library, as outlined in **Figure 8**. To add an item to Favorites, touch and hold it in the library. The Item menu appears, as outlined in **Figure 9**. Touch **Add to Favorites**. The item is added to the Favorites. Touch and hold an item in the Favorites and touch **Remove from Favorites**. The item is removed from the Favorites.

*Note: Touch the screen and move your finger up to view more Favorites when there are more than four in the list.*

*Figure 8: Favorites in the Library*

*Figure 9: Item Menu*

# 7. Connecting the Kindle Fire to a PC or Mac

eBooks and other files that you have obtained elsewhere can be imported to the Kindle Fire. To import media:

1. Connect the Kindle Fire to your computer using the provided USB cable. The USB Connected screen appears, as shown in **Figure 10**.
2. Open **My Computer** on a PC and double-click the 'KINDLE' removable drive or double-

   click the icon on a Mac. The Kindle Folders open on a PC, as shown in **Figure 11**, or on a Mac, as shown in **Figure 12**.
3. Double-click a folder. The folder opens.
4. Drag and drop a file into the open folder. The file is copied and will appear in the corresponding library.

*Figure 10: USB Connected Screen*

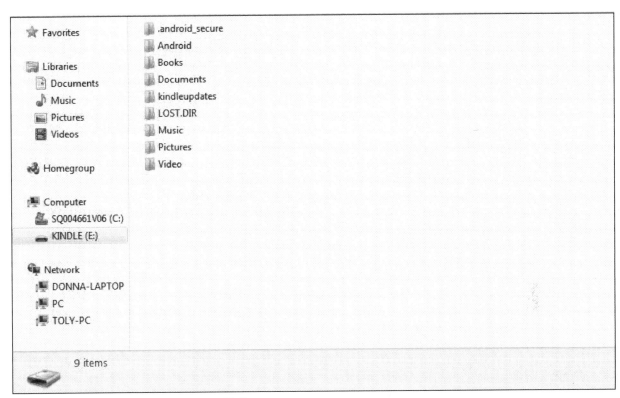

*Figure 11: Kindle Folders on a PC*

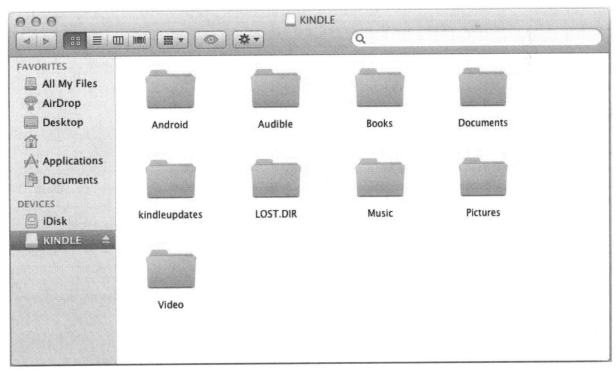

*Figure 12: Kindle Folders on a Mac*

# Managing eBooks and Periodicals

---

## Table of Contents

---

## 1. Buying an eBook on the Kindle Fire

You can buy an eBook from the Amazon Kindle Store using your Kindle Fire. To buy an eBook:

***Warning: Before touching BUY, make sure you want the eBook. The Kindle Store on the Kindle uses one-click purchasing. Once you leave the Confirmation screen, you cannot cancel the order.***

1. Touch **Books** at the top of the Library. The Books Library appears, as shown in **Figure 1**.
2. Touch **Store** at the top right of the screen. The Book Store opens, as shown in **Figure 2**.
3. Touch **Search Book store** at the top of the screen. The virtual keyboard appears at the bottom of the screen.
4. Type the name of an author or eBook and touch the Go button. A list of available eBook results appears, as shown in **Figure 3**.
5. Touch the title of an eBook. The eBook description appears, as shown in **Figure 4**.
6. Touch **Buy for $##.##** where the ##.## represents the price of the eBook. The eBook is purchased and downloaded to your Kindle Fire Library.

*Note: Touch* **Cancel Order** *below the* Read Now *button on the Confirmation screen if you did not mean to purchase the eBook.*

*Figure 1: Books Library*

*Figure 2: Book Store*

*Figure 3: List of Available eBook Results*

*Figure 4: eBook Description*

## 2. Buying or Subscribing to a Periodical

You can buy or subscribe to a newspaper or magazine from the Amazon Kindle Store using your Kindle Fire. To buy or subscribe to a periodical:

***Warning: Before touching the*** Buy current issue ***button or the*** Subscribe now ***button, make sure you want the periodical issue. The Kindle Store on the Kindle uses one-click purchasing. Unlike with eBook orders, you will not be given an opportunity to cancel a periodical order from the Confirmation screen.***

1. Touch **Newsstand** at the top of the Library. The Newsstand Library appears, as shown in **Figure 5**.
2. Touch **Store** at the top right of the screen. The Newsstand Store opens, as shown in **Figure 6**.
3. Touch **Search Newsstand store** at the top of the screen. The virtual keyboard appears at the bottom of the screen.
4. Type the name of a periodical and touch the Go button. A list of available periodical results appears, as shown in **Figure 7**.
5. Touch the title of a newspaper. The Periodical description appears, as shown in **Figure 8**.
6. Touch the Buy current issue button to purchase an issue or touch the Subscribe now button to subscribe. A confirmation is shown and the periodical issue appears in the library. You have 14 days to cancel your subscription before you are charged for the first time. Refer to *"Cancelling Your Newspaper or Magazine Free Trial"* on page 32 to learn how to cancel your subscription.

*Figure 5: Newsstand Library*

*Figure 6: Newsstand Store*

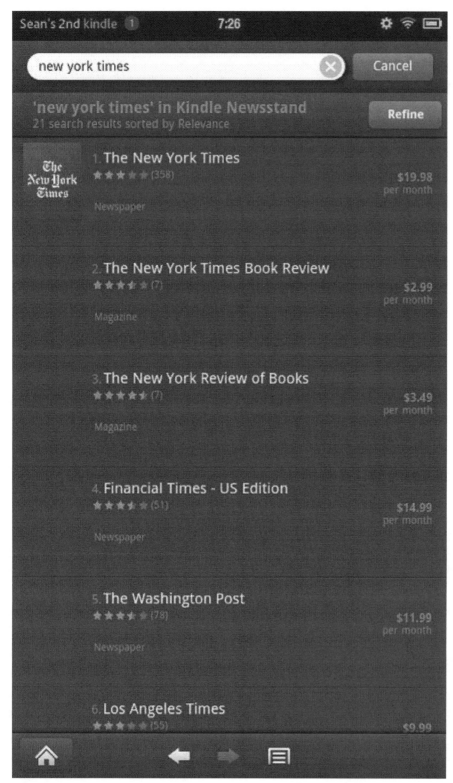

*Figure 7: List of Available Periodical Results*

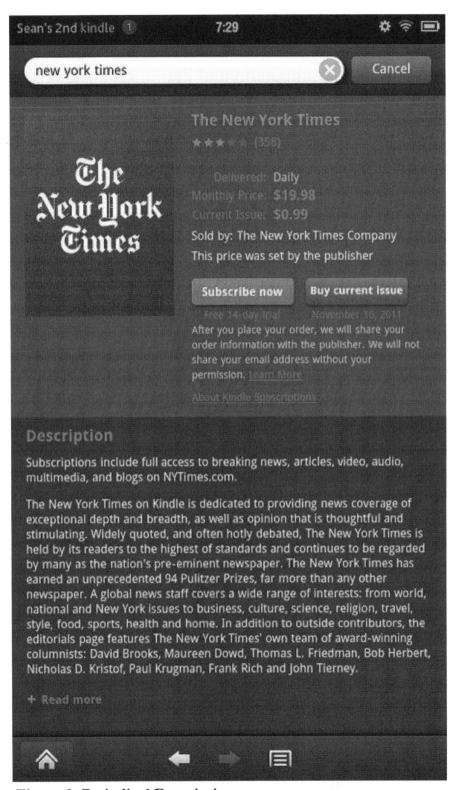

*Figure 8: Periodical Description*

# 3. Cancelling Your Newspaper or Magazine Free Trial

To cancel a subscription to a newspaper or magazine, use the Amazon website on your computer or Kindle Fire. To cancel a subscription:

1. Go to **www.amazon.com/myk** using your computer's internet browser or the Amazon Silk browser on the Kindle Fire. The Kindle Information screen appears, as shown in **Figure 9**.
2. Click **Subscription Settings**, outlined on the left side of the screen in **Figure 9**. The Subscription Settings screen appears, as shown in **Figure 10**. Your active subscriptions are shown on this screen.
3. Click **Actions** next to the subscription you wish to cancel. The Subscription options appear.
4. Click **Cancel Subscription**. A confirmation window appears.
5. Click the **Cancel Subscription** button. The subscription is cancelled.

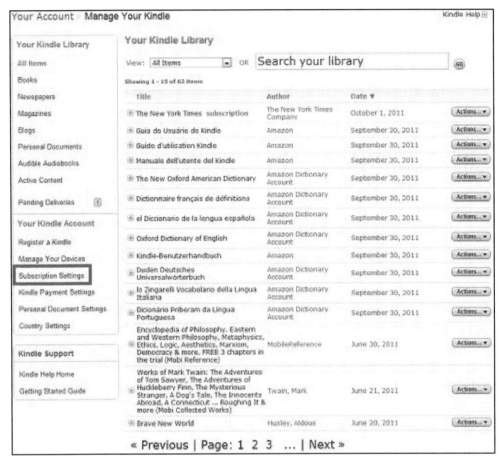

*Figure 9: Kindle Information Screen*

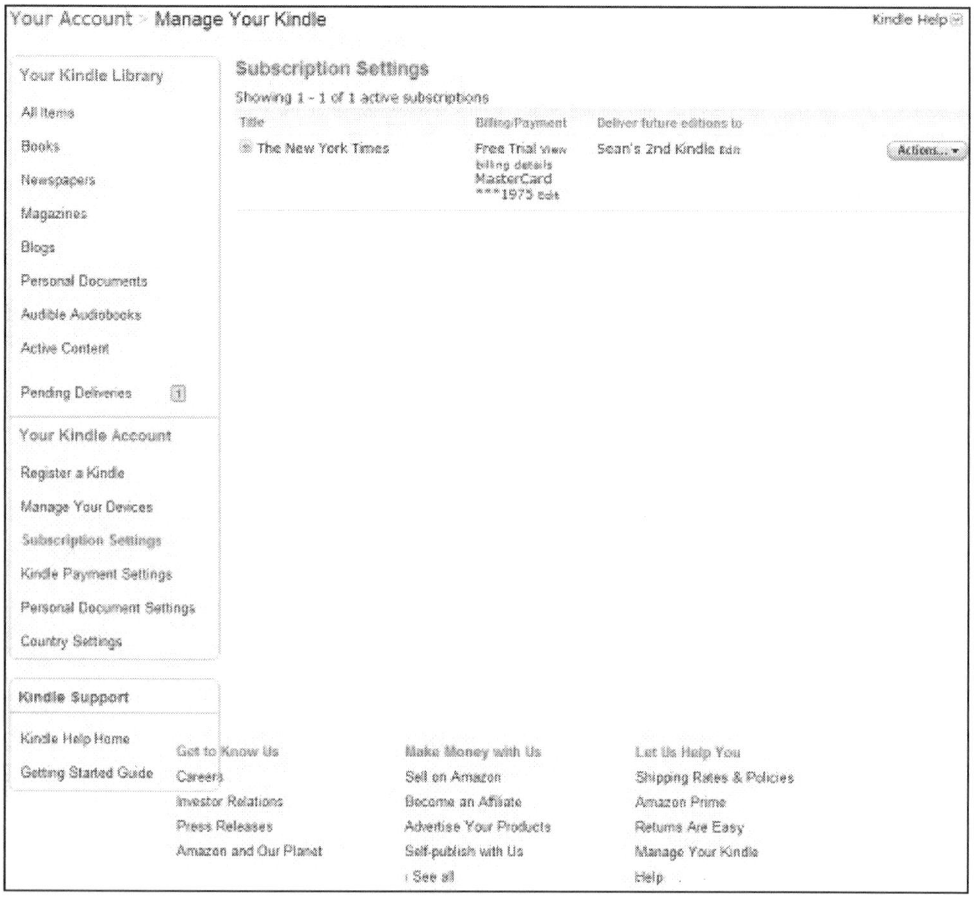

*Figure 10: Subscription Settings Screen*

# 4. Browsing Recommendations

Amazon makes recommendations based on the eBooks you have viewed or purchased. To view these recommendations:

*Note: Only the Book Store offers customized recommendations. The Newsstand Store does not have this feature.*

1. Touch **Books** at the top of the Library. The Books Library appears.
2. Touch **Store** at the top right of the screen. The Book Store opens.
3. Touch **See all** next to 'Recommended for You'. A list of recommendations appears, as shown in **Figure 11**.

*Note: Refer to "Buying an eBook on the Kindle Fire"* *on page 22 to learn how to purchase an eBook.*

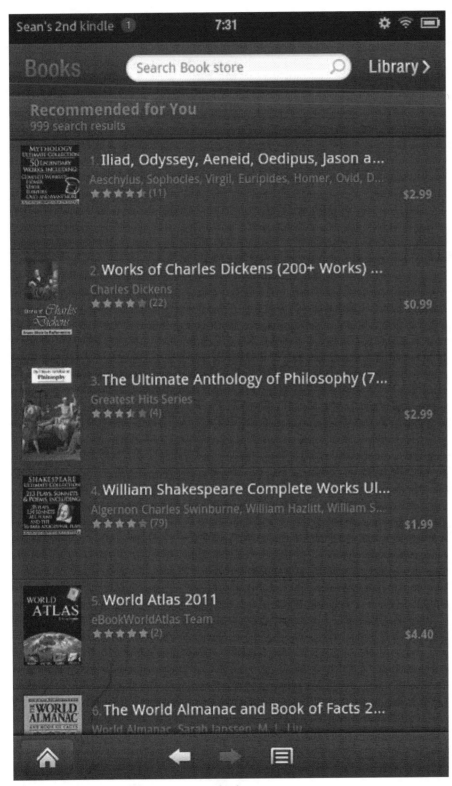

*Figure 11: List of Recommendations*

# 5. Buying an eBook on Amazon.com Using Your Computer

In addition to using the Kindle Fire, eBooks can be purchased online on Amazon.com using your PC or Mac and then transferred to your tablet. Refer to *"Connecting the Kindle Fire to a PC or Mac"* on page 19 to learn more. To search for and purchase an eBook on Amazon.com:

***Warning: Before clicking 'Buy now with 1-Click', make sure you want the eBook. The Kindle Store on Amazon.com uses one-click purchasing. Once you leave the Confirmation screen, you cannot cancel the order.***

1. Go to **www.amazon.com** using your computer's Web browser.
2. Point your mouse cursor at **Shop by Department**. A list of departments appears.
3. Point your mouse cursor at **Books**. The Books category appears, as outlined in **Figure 12**.
4. Click **Kindle Books**. The Kindle Store opens and the eBook categories appear on the left side of the screen, as shown in **Figure 13**.
5. Click a genre in the Books menu on the left side of the screen. Keep clicking genres on the left side on the following screens. A list of eBooks is shown each time. Search for a specific eBook or author by clicking on the Search drop-down menu.
6. Click the eBook you wish to purchase. The eBook Description screen appears.
7. Select the name of your Kindle from the 'Deliver to' drop-down menu. Your Kindle is selected.
8. Click **Buy now with 1-Click**. A Confirmation screen appears and the item is delivered to your Kindle.

*Note: A purchased eBook is only delivered to the device selected in step 7. To download the eBook to another registered device, open the Archived items and restore it. Refer to* "Archiving an eBook or Periodical" *on page 38 to learn how to restore an eBook on the Kindle Fire.*

*Figure 12: Books Category on Amazon.com*

*Figure 13: Kindle Store*

# 6. Archiving an eBook or Periodical

An eBook or periodical can be removed from your Kindle and placed in the Amazon Cloud where it does not take up space on your device. An archived eBook is retrievable using the wireless connection. To archive an eBook or periodical, touch and hold the cover. The Item menu appears, as outlined in **Figure 14**. Touch **Remove from Device**. The eBook or periodical is archived in the Amazon Cloud.

*Note: To restore an archived eBook, touch **Cloud** in the Books library and then touch the book cover.*

*Figure 14: Item Menu*

# 7. Keeping a Periodical Issue when a New Issue is Downloaded

When you have a subscription to a periodical, you can have the Kindle Fire either delete the previous monthly issue every time a new one arrives or keep the issue. To retain the previous periodical issue, touch and hold the cover. The Item menu appears. Touch **Keep**. The issue will be kept when a new one is downloaded. To automatically delete the current issue, touch and hold the cover again. The Item menu appears. Touch **Don't Keep**. The issue will be deleted when a new one is downloaded.

*Note: By default, the Kindle Fire will delete the previous issue when a new one is downloaded.*

# 8. MobileReference eBooks

The quality of Kindle eBooks varies greatly between publishers. MobileReference only publishes books that are carefully checked for accuracy and completeness by a team of experts.

**The MobileReference Active Table of Contents**

Looking for a poem or a story on an electronic device in an eBook that contains hundreds of stories is greatly aided by an Active Table of Contents. The Active Table of Contents provides quick access to eBook contents via hyperlinks. Automatically scanned eBooks lack an Active Table of Contents and links to footnotes. MobileReference editors carefully examine each eBook and insert links to individual stories, poems, letters, and footnotes to improve the reading experience and simplify electronic eBook navigation. The result is high quality books with an Active Table of Contents. **Figure 15** shows an example of an Active Table of Contents on the Kindle Fire.

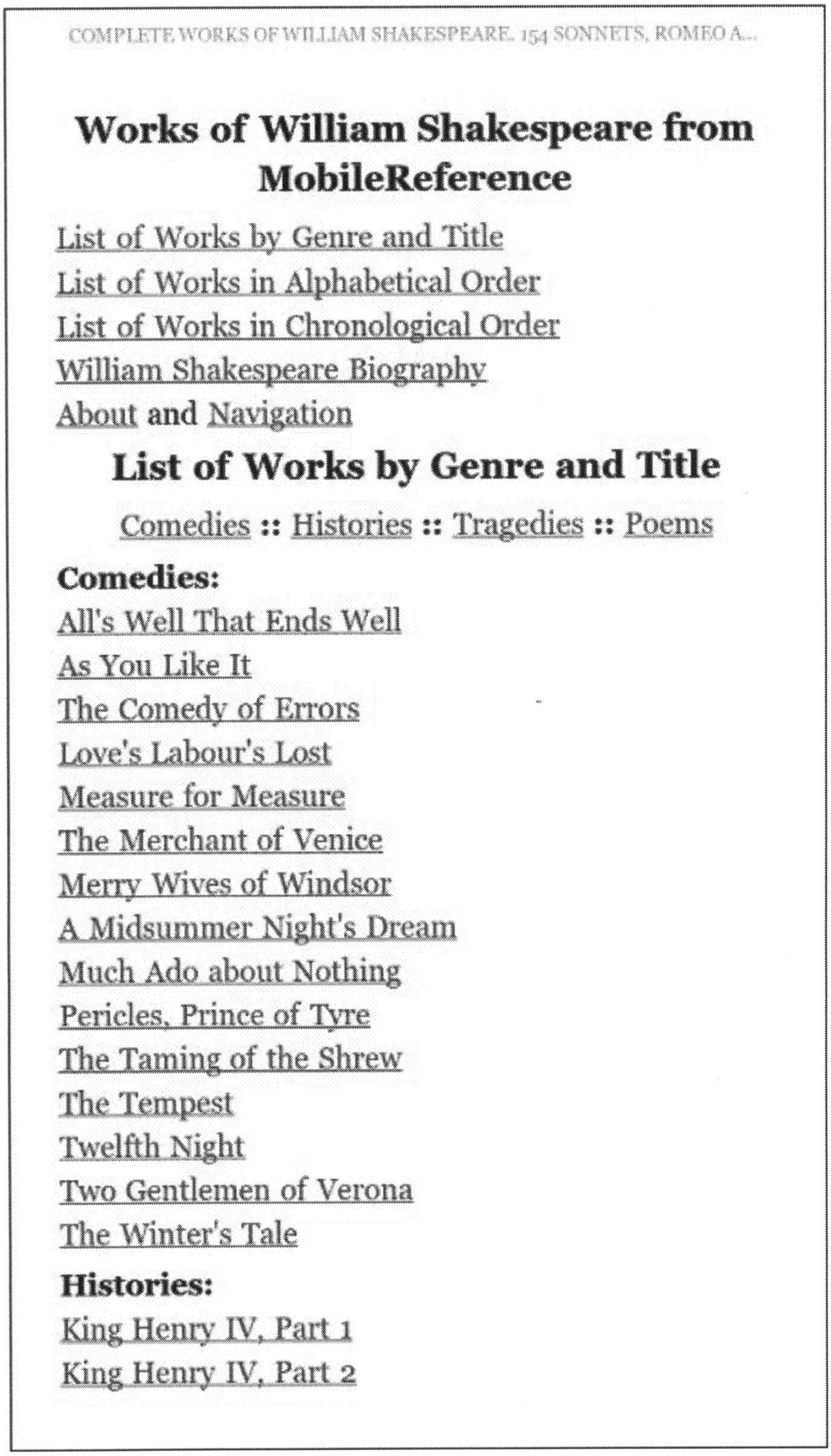

COMPLETE WORKS OF WILLIAM SHAKESPEARE. 154 SONNETS, ROMEO A...

**Works of William Shakespeare from MobileReference**

List of Works by Genre and Title
List of Works in Alphabetical Order
List of Works in Chronological Order
William Shakespeare Biography
About and Navigation

**List of Works by Genre and Title**

Comedies :: Histories :: Tragedies :: Poems

**Comedies:**

All's Well That Ends Well
As You Like It
The Comedy of Errors
Love's Labour's Lost
Measure for Measure
The Merchant of Venice
Merry Wives of Windsor
A Midsummer Night's Dream
Much Ado about Nothing
Pericles, Prince of Tyre
The Taming of the Shrew
The Tempest
Twelfth Night
Two Gentlemen of Verona
The Winter's Tale

**Histories:**

King Henry IV, Part 1
King Henry IV, Part 2

*Figure 15: Active Table of Contents on the Kindle Fire*

**MobileReference Author Collections**

MobileReference also pioneered the publication of author-collected works in one eBook. This was motivated in part by an effort to reduce the clutter of titles in a digital library. Consider the 'Works of Charles Dickens', which contains over 200 works. When purchased individually, each eBook has an entry in your digital library (analogous to a spine of a paper book). To select an eBook in a digital library of 200 books, one needs to scroll through as many as 20 pages of the library (when ten books are presented on each page). MobileReference author collections organize all eBooks into one electronic volume that has a single representation in a digital library: 'Works of Charles Dickens'.

Inside author collections, MobileReference editors diligently create categorical, alphabetical, and chronological eBook indices. For example, a reader looking for 'A Christmas Carol' can click Fiction > 'A Christmas Carol'. Alternatively, a reader can click 'List of works in alphabetical order' > 'C' > 'A Christmas Carol'. If a reader forgets the book title but remembers that 'A Christmas Carol' was written by Charles Dickens early in his career, he or she can click on the 'List of works in chronological order' > (1843) 'A Christmas Carol'.

By 2011, MobileReference had developed over two hundred collections by Shakespeare, Jane Austen, Mark Twain, Conan Doyle, Jules Verne, Dickens, Tolstoy, and other authors. If you do not see the eBook you want, please email us: support@soundtells.com (MobileReference is an imprint of SoundTells, LLC).

MobileReference author collections cost $5.99 or less.

To search for MobileReference eBooks: enter mobi (shortened MobileReference) and a keyword; for example: mobi Dickens

# Finding Free eBooks

Here is a list of websites offering free eBooks. Please note that only PRC and PDF eBooks can be added to the Kindle Fire library. Refer to *"Connecting the Kindle Fire to a PC or Mac"* on page 19 to learn how to transfer downloaded eBooks to your Kindle. In addition, HTML eBooks can be viewed using the Kindle Fire's Web browser. Refer to *"Using the Silk Web Browser"* on page 136 to learn how. For other file types, please use a computer or other eReader to view the eBook.

## Table of Contents

# 1. Arthur's Classic Novels

**Web address**: http://arthurclassicnovels.com

**Number of available books**: 4000
**Categories**: 20th Century classics, Philosophy, Religion, History, Technology, Mystery, Children's, and many more
**Prominent authors**: Jules Verne, Charles Dickens, Friedrich Nietzsche, Ralph Waldo Emerson, Fyodor Dostoyevsky, and more
**Available formats**: HTML
**How to use**: Search by title or author or browse by category. Click on a title to view it on a separate web page. Use the next and previous page buttons to scroll by page.

**Additional Tips**: There are no eBooks available for download from this website. However, to read an eBook, wireless is not required. Load up the web page containing the eBook and turn wireless off to preserve battery life.

# 2. Baen Free Library

**Web address**: http://www.baen.com/library

**Number of available books**: 100
**Categories**: Science fiction
**Prominent authors**: John Joseph Adams, David Friedman, Richard Roach, and more
**Available formats**: Kindle/Mobi/Palm (PRC), Nook/Stanza (ePub), Microsoft Reader, Sony Digital, RTF, and HTML
**How to use**: Browse by series, authors, or titles by clicking an option on the left side of the screen. Click an eBook to view its description and then download a zipped or unzipped file. Click 'Read Online' to view the HTML version of the book. Click 'Email book to my Kindle' to send the eBook to your Kindle email. The charge is 15 cents per each megabyte. For example, emailing a ten-megabyte eBook to the Kindle costs $1.50.
**Additional Tips**: Several reader programs are available for download on Baen. Download the Microsoft Reader, Mobipocket Reader, and Rocket eBook here.

## 3. BeBook Catalog

**Web address**: http://mybebook.com/download_free_ebook

**Number of available books**: 20,000
**Categories**: Various
**Prominent authors**: Leonardo da Vinci, Sunzi, Arthur Conan Doyle, Jane Austen, William Shakespeare, and more
**Available formats**: Plain text (TXT), PDF
**How to use**: Click a letter to browse by author or title. Click a title to view its description. Click 'download' next to 'Plain text' or 'PDF' to download an eBook. Most eReaders and other devices can open PDF files.

**Additional Tips**: To browse books in another language, click the arrow next to 'Please select language'. After choosing a language, a list of the available books in that language appears.

## 4. Bookyards Library

**Web address**: http://www.bookyards.com

**Number of available books**: 17,000
**Categories**: Classics
**Prominent authors**: Geoffrey Chaucer, Confucius, Robert Burns, Hans Christian Andersen, Julius Caesar, and more
**Available formats**: PDF
**How to use**: Click a letter to browse by title, author, or biography. Click a title to view the eBook description. Click 'Download' to download the eBook.
**Additional Tips**: Many external links are provided under each author. The extra websites are excellent resources for additional free eBooks.

# 5. Christian Classics Ethereal Library

**Web address**: http://www.ccel.org

**Number of available books**: 1200
**Categories**: Christian
**Prominent authors**: Various
**Available formats**: Plain text (TXT), HTML, PDF ($2.95)
**How to use**: Click 'Browse' to peruse the authors, titles, subjects, and languages. Click an author to view the available titles. Click a title to view its description. Click one of the formats to download the eBook.
**Additional Tips**: This website has its own online reader, which is similar to an electronic reader. Click the hyperlinks in each eBook to navigate to another excerpt. Books are available in 13 languages.

# 6. E-Books Directory

**Web address**: http://www.e-booksdirectory.com

**Number of available books**: 4600
**Categories**: Educational, Business, Engineering, Law, Medicine, Nature, Travel
**Prominent authors**: Various
**Available formats**: Various (see Additional Tips)
**How to use**: Click a genre. A list of related books appears. Click a title. The title description appears. Click 'Download' or 'Read' to access the book.
**Additional Tips**: Each title on this website links to an external source where the eBook can be read or downloaded. Therefore, the number of formats and types of books varies greatly. This website is excellent as a reference tool and a starting point for finding educational works.

## 7. Elegant Solutions Software and Publishing Company: eBooks for people who think

**Web address**: http://esspc-ebooks.com

**Number of available books**: 444

**Categories**: Modern Fiction, Classics, Children's, Romance, History, many more

**Prominent authors**: H.B. Irving, Mark Twain, Henry James, Bram Stoker, The Brothers Grimm, and more

**Available formats**: Kindle/Mobi/Palm (PRC), Microsoft Reader (LIT)

**How to use**: Click 'New eBooks' or 'All titles' at the top of the page to browse books. Author, Title, and Genre searches are also available. Click the Microsoft Reader icon to download the LIT version of the book. Click the Mobipocket icon to download the PRC version of the book. Additional Tips: Links are provided to download the Microsoft Reader as well as other useful software.

## 8. Feedbooks

**Web address**: http://www.feedbooks.com

**Number of available books**: 5300

**Categories**: Various, Mostly Short Stories and Science Fiction

**Prominent authors**: Francis Scott Fitzgerald, Edgar Allan Poe, Kurt Vonnegut, Isaac Asimov, Oscar Wilde, and more

**Available formats**: Nook/Stanza (ePub), Kindle/Mobi/Palm (PRC), and PDF

**How to use**: On the main page, click 'Public Domain' to view all free eBooks. Click a category on the left side of the page to browse eBooks. Click a title to view its description. Click a format to download the eBook.

**Additional Tips**: Feedbooks is an online community for bibliophiles. Click 'Register' at the top right of the page to create an account. With a registered account, books can be added to a favorites list, comments can be posted to eBook descriptions, and PDF's with custom settings can be downloaded.

# 9. Girlebooks - eBooks by female authors

**Web address**: http://girlebooks.com

**Number of available books**: 137
**Categories**: Female authors only
**Prominent authors**: Jane Austen, Louisa May Alcott, Virginia Woolf, Mary Shelley, Kate Chopin, and more
**Available formats**: Nook/Stanza (ePUB), Kindle/Mobi/Palm (PRC), Microsoft Reader (LIT), Plain Text (TXT), PDF
**How to use**: Click 'eBook Catalog' to view all eBooks. Click 'Select Category' to choose a genre. Click an eBook cover to view its description. Click a format to download it.

**Additional Tips**: Use the 'Also available at:' links to find the eBook on another website. Use the 'For Authors' link to submit your own work for consideration.

# 10. Project Gutenberg

**Web address**: http://www.gutenberg.org/wiki/Main_Page

**Number of available books**: 33,000
**Categories**: Classics
**Prominent authors**: Leo Tolstoy, Charles Dickens, Homer, Voltaire, Mary Shelley, and more
**Available formats**: Nook/Stanza (ePUB), Kindle/Mobi/Palm (PRC), HTML, Plucker (PDB), Plain Text (TXT), QiOO Mobile (QIOO)
**How to use**: Click 'Browse Catalog' and then click a letter under 'Authors' or 'Titles'. Click a language to view all related books. Click a title to view its description. Click the 'Download' tab to view all available formats. Click a format to download the eBook.

**Additional Tips**: Click 'Read this eBook online' to view the online HTML version of the book. Click 'Bookshelf' on the home page to view eBook suggestions for various genres. Click 'Partners, Affiliates, and Resources' on the home page to view links to additional Gutenberg sites and other eBook resources.

# 11. Internet Text Archive

**Web address**: http://www.archive.org/details/texts

**Number of available books**: 2,000,000
**Categories**: Various
**Prominent authors**: Various
**Available formats**: Various
**How to use**: This website is a reference tool for finding free eBooks on the internet. Use the Search field at the top of the page to find eBooks. Click the drop-down menu to select a database to search.

**Additional Tips**: Audio books are also available. Many links are provided to educational and other useful websites.

# 12. ManyBooks.net

**Web address**: http://manybooks.net

**Number of available books**: 29,000
**Categories**: Various
**Prominent authors**: Winston Churchill, T.S. Eliot, William Shakespeare, Karl Marx, Napoleon, and more
**Available formats**: Nook/Stanza (ePUB), Kindle/Mobi/Palm (PRC), HTML, PalmDOC (PDB), Plain Text (TXT), PDF, Rich Text (RTF), Sony Reader (LRF)
**How to use**: Click 'Authors', 'Titles', 'Genres', or 'Languages' to browse books. Click a title to view its description. Click the 'Download' drop-down menu to select a format. Click 'Download' to download the eBook. Some formats require registration.

**Additional Tips**: Registering with the website gives access to the Wishlist feature. Wikipedia and WorldCat links are provided to give more information about authors.

# 13. MobileRead Uploads

**Web address**: http://www.mobileread.com/forums/ebooks.php?order=desc&sort=dateline

**Number of available books**: 16,000
**Categories**: Action, Biography, Spiritual, Philosophy, Young Adult
**Prominent authors**: Daniel Defoe, H.G. Wells, Hans Christian Andersen, Agatha Christie, Bertrand Russell, and many more
**Available formats**: Nook/Stanza (ePUB), Kindle/Mobi/Palm (PRC), Sony Reader (LRF), eBookwise (IMP)
**How to use**: Click the 'Format' drop-down menu to select a format to display. Select 'ALL' to see all eBook formats. Click the 'Genre' drop-down menu to select an eBook category. Click 'Go' to view all related media. Click a title to view its description. Scroll down to the bottom of the page and click one of the attached files to download the eBook.

**Additional Tips**: MobileRead is an eBook community that contains blogs about many eBook topics. Links to other useful resources are also provided.

# 14. Munseys

**Web address**: http://www.munseys.com

**Number of available books**: 20,000
**Categories**: Various
**Prominent authors**: Francis Bacon, Lord Byron, O'Henry, Robert Sheckley, Ivan Turgenev, and more
**Available formats**: Nook/Stanza (ePUB), Kindle/Mobi/Palm (PRC), HTML, Plucker (PDB), PDF, Sony Reader (LRF), and Microsoft Reader
**How to use**: Scroll down on the home page to browse by category. Click the 'By' drop-down menu to select a search method and type in the related search term. Click a letter at the top of the search results to display all authors whose last name starts with that letter.

**Additional Tips**: Munseys uses tags, such as genres, titles, authors, and more. Tags allow readers to find books more easily. There is also a blog for sharing thoughts with fellow readers.

# 15. Planet eBook

**Web address**: http://www.planetebook.com

**Number of available books**: 82
**Categories**: Classics
**Prominent authors**: George Orwell, Jonathan Swift, John Milton, Franz Kafka, Herman Melville, and more
**Available formats**: PDF
**How to use**: All eBooks are listed on the home page. Click a title to view its description. Right-click '1-page version' (portrait) or '2-page version' (landscape) and click 'Save link as'. Click 'Save' to download the eBook.
**Additional Tips**: Blog and newsletter are available on this website.

# 16. Planet PDF

**Web address**: http://www.planetpdf.com/free_pdf_ebooks.asp

**Number of available books**: 60
**Categories**: Classics
**Prominent authors**: Charles Dickens, Aesop, James Joyce, Edgar Allan Poe, Robert Louis Stevenson, and more
**Available formats**: PDF
**How to use**: All available eBooks are listed on the home page. Click a title to view its description. Right-click a PDF icon and click 'Save link as'. Click 'Save' to download the eBook.

**Additional Tips**: This website provides tagged PDFs, which are optimized for eReaders, allowing better navigation on your Kindle.

# 17. Project Runeberg - Nordic Literature

**Web address**: http://runeberg.org

**Number of available books**: 1067
**Categories**: Nordic
**Prominent authors**: Various
**Available formats**: HTML
**How to use**: Click 'Catalog' at the top of the page to view all titles. Click a title to read it.

**Additional Tips**: The first eBook in the Catalog is a guide to learning the Norsk (Icelandic) language. This is necessary to be able to read the books.

# 18. Stanford Collection

**Web address**: http://collections.stanford.edu

**Number of available books**: Not applicable
**Categories**: Educational
**Prominent authors**: Various
**Available formats**: Various
**How to use**: Click a link to an external database to browse it.
**Additional Tips**: Not applicable

# 19. World Wide School

**Web address**: http://www.worldwideschool.org/library/catalogs/bysubject-top.html

**Categories**: Classics, Youth, History, Technology, Philosophy, Religion
**Prominent authors**: Various
**Available formats**: HTML
**How to use**: Click a genre to browse the related books. Click a title to read it. Click 'By Title' or 'By Author' at the top to sort the books accordingly.

**Additional Tips**: Not applicable

# Reading eBooks and Periodicals

## Table of Contents

# 1. Navigating an eBook or Periodical

The Kindle Fire makes it easy to navigate an eBook or periodical. Use the following tips while reading:

- **Navigating the Pages** - Touch the screen and move your finger to the left to turn to the next page or to the right to turn to the previous one.

- **Navigating to the Table of Contents** - Touch the screen anywhere (as long as it is not a link). The eBook menu appears at the bottom of the screen, as shown in **Figure 1**. Touch the ▦ button and then touch **Table of Contents**.

- **Navigating to the Beginning of an eBook or Periodical** - Touch the screen anywhere (as long as it is not a link). The eBook menu appears. Touch the ▦ button and then touch **Beginning**.

- **Navigating to a Specific Location** - Touch the screen anywhere (as long as it is not a link). The eBook menu appears. Touch the ⬤ on the ━━━━━⬤━━━━━ bar and drag it to the left or right to select a specific location in the eBook or periodical.

- **Sync to Furthest Page Read** - Touch the screen anywhere (as long as it is not a link). The eBook menu appears. Touch the ▦ button and then touch **Sync to Furthest Page** to open the furthest read page on all devices registered to the same Amazon account as the Kindle Fire.

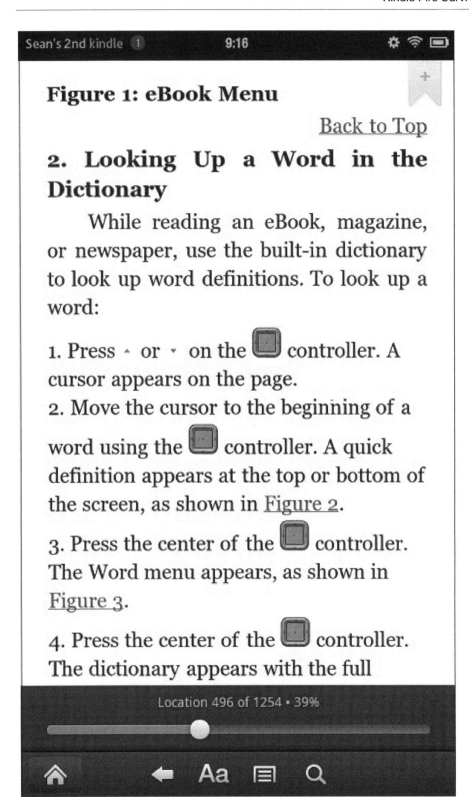

*Figure 1: eBook Menu*

## 2. Looking Up a Word in the Dictionary

While reading an eBook or periodical, use the built-in dictionary to look up word definitions. To look up a word, touch and hold it. A quick definition appears, as shown in **Figure 2**. Touch **Full Definition**. The full dictionary definition appears, as shown in **Figure 3**. Touch anywhere on the screen and touch the button to resume reading where you left off.

KINDLE 2011 SURVIVAL GUIDE: STEP-BY-STEP USER GUIDE FOR KINDLE 2...

**Figure 6: Note Window**

Back to Top

## 5. Adding a Bookmark

While reading an eBook or periodical, the media can be bookmarked at certain locations. To add a bookmark

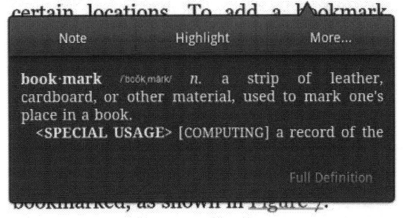

bookmarked, as shown in Figure 7.

*Note: Refer to* Viewing Your Notes, Highlights, and Bookmarks *to learn how to view your list of bookmarks.*

*Figure 2: Quick Definition*

THE NEW OXFORD AMERICAN DICTIONARY

**book·mark** /ˈboŏkˌmärk/ *n.* a strip of leather, cardboard, or other material, used to mark one's place in a book.

   **<SPECIAL USAGE>** [COMPUTING] a record of the address of a file, web page, or other data used to enable quick access by a user.

■ *v.* [COMPUTING] record the address of (a file, web page, or other data) for quick access by a user: *if you think politics is the ultimate game, be sure to bookmark eVote.*

---

**book·mo·bile** /ˈboŏkməˌbēl/ *n.* a truck, van, or trailer serving as a mobile library.

**<ORIGIN>** 1930s: from BOOK, on the pattern of *automobile.*

*Linked entries:*
BOOK ■

---

**Book of Chang·es** *n.* another name for I CHING.

*Figure 3: Full Dictionary Definition*

## 3. Highlighting a Word or Phrase

While reading an eBook or periodical, words and phrases can be highlighted. To highlight a word or phrase, touch and hold a single word until the magnifying glass appears, as shown in **Figure 4**. Without letting go of the screen, drag your finger to select a phrase. The phrase is selected. Touch **Highlight**. The phrase is highlighted., as shown in **Figure 5**. You can also highlight a single word by letting go of the screen as soon as you see the magnifying glass and touching **Highlight**.

*Note: Refer to* "Viewing Your Notes, Highlights, and Bookmarks" *on page 64 to learn how to view your list of highlights.*

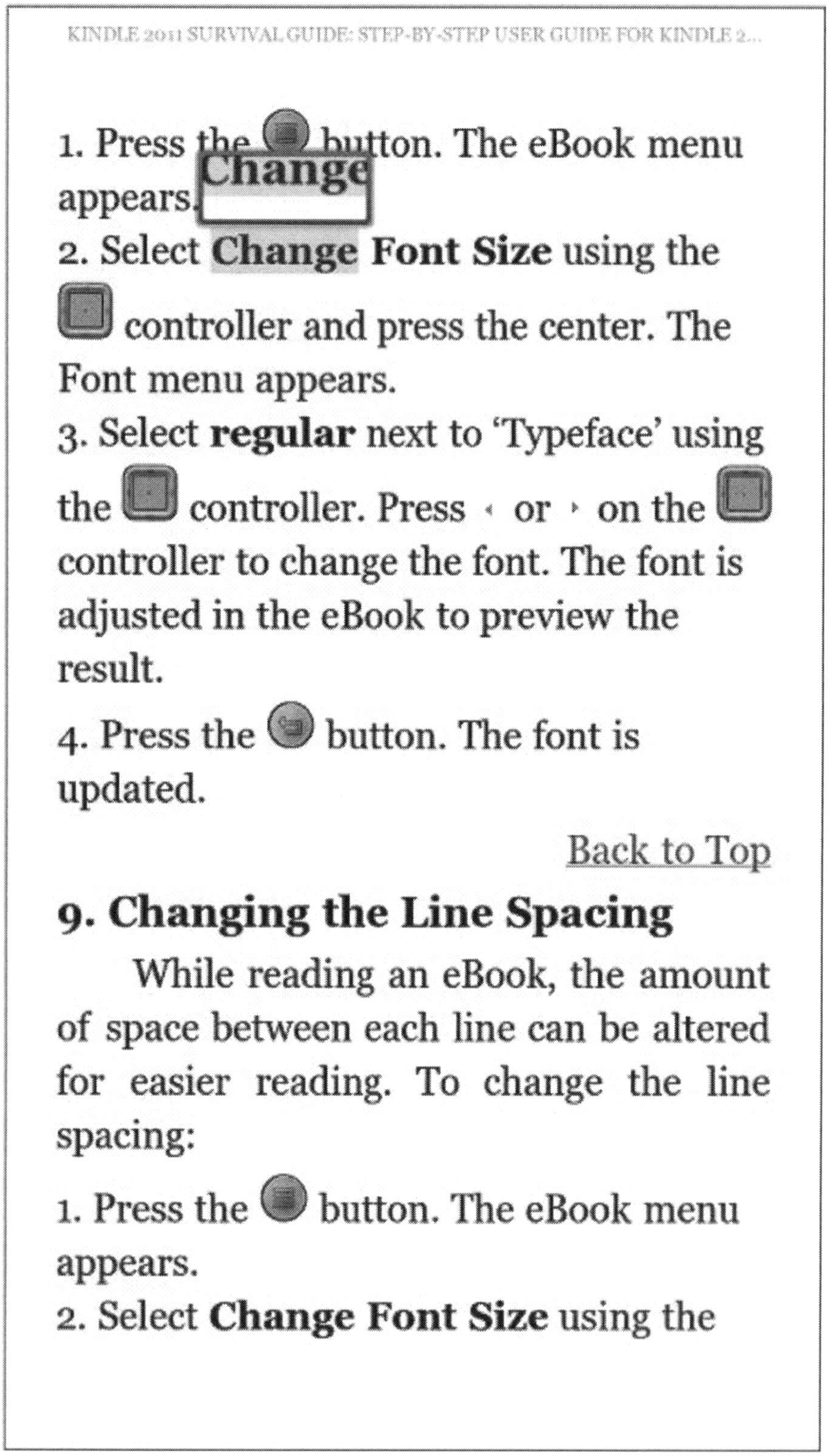

1. Press the ⬤ button. The eBook menu appears. **Change**

2. Select **Change** **Font Size** using the ⬛ controller and press the center. The Font menu appears.

3. Select **regular** next to 'Typeface' using the ⬛ controller. Press ‹ or › on the ⬛ controller to change the font. The font is adjusted in the eBook to preview the result.

4. Press the ⬤ button. The font is updated.

Back to Top

## 9. Changing the Line Spacing

While reading an eBook, the amount of space between each line can be altered for easier reading. To change the line spacing:

1. Press the ⬤ button. The eBook menu appears.

2. Select **Change Font Size** using the

*Figure 4: Magnifying Glass*

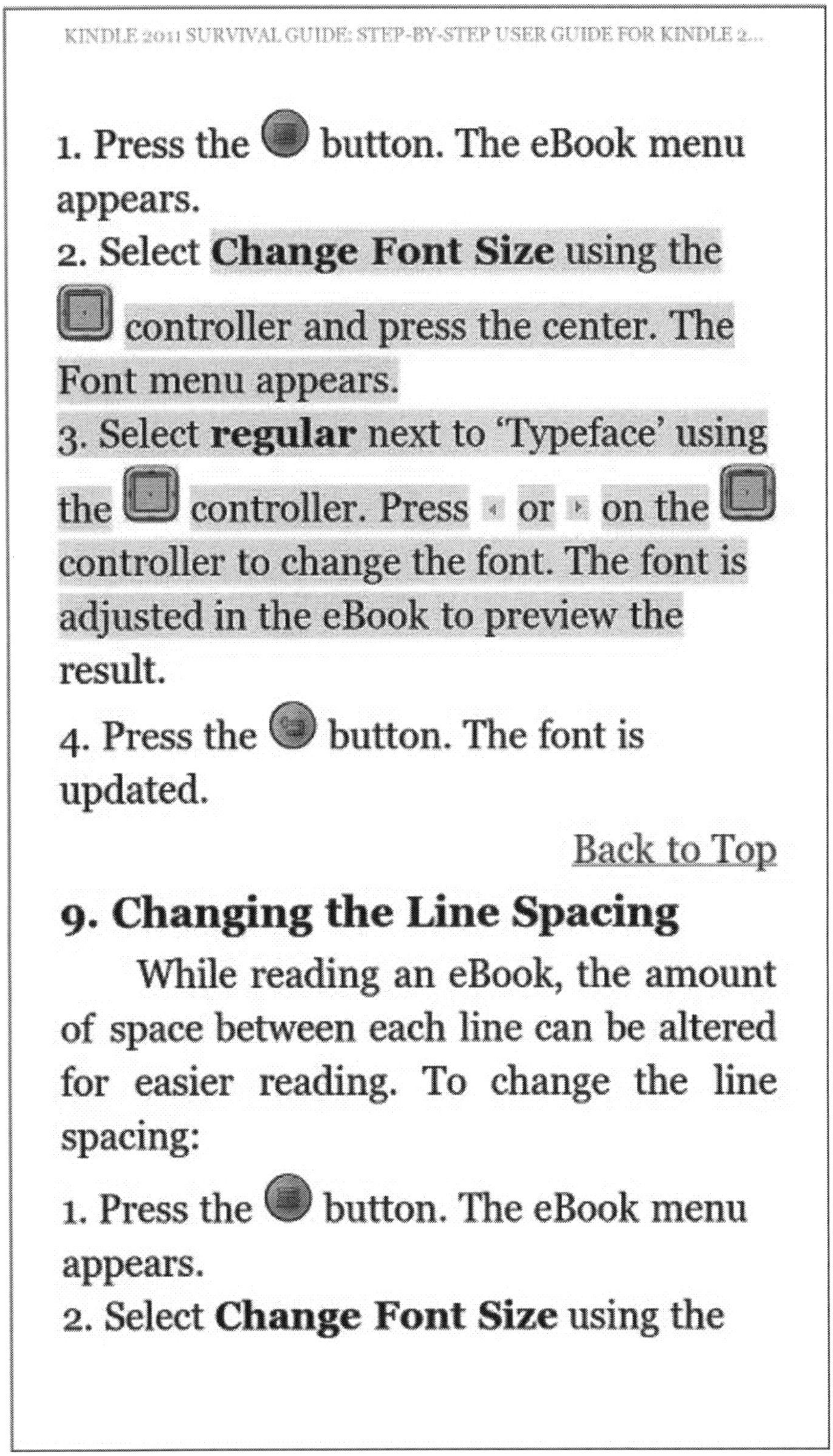

1. Press the ⬤ button. The eBook menu appears.

2. Select **Change Font Size** using the ▣ controller and press the center. The Font menu appears.

3. Select **regular** next to 'Typeface' using the ▣ controller. Press ◂ or ▸ on the ▣ controller to change the font. The font is adjusted in the eBook to preview the result.

4. Press the ⬤ button. The font is updated.

Back to Top

## 9. Changing the Line Spacing

While reading an eBook, the amount of space between each line can be altered for easier reading. To change the line spacing:

1. Press the ⬤ button. The eBook menu appears.

2. Select **Change Font Size** using the

*Figure 5: Highlighted Phrase*

# 4. Making a Note

While reading an eBook or periodical, notes can be added. To add a note, touch and hold a single word until the magnifying glass appears. Release the screen. Touch **Note**. The virtual keyboard appears at the bottom of the screen. Type a note and touch the button. A note is added and a icon appears next to the word.

*Note: Refer to* "Viewing Your Notes, Highlights, and Bookmarks" *on page 64 to learn how to view your list of notes.*

# 5. Adding a Bookmark

While reading an eBook or periodical, the media can be bookmarked in order to quickly find the same location in the future. To add a bookmark, touch the top right corner of the screen. A appears at the top right corner of the screen to indicate that the page is bookmarked, as shown in **Figure 6**.

*Note: Refer to* "Viewing Your Notes, Highlights, and Bookmarks" *on page 64 to learn how to view your list of bookmarks.*

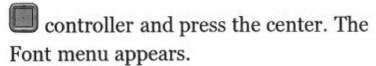 controller and press the center. The Font menu appears.

3. Select **medium** next to 'Line Spacing' using the ⬜ controller. Press ‹ or › on the ⬜ controller to change the line spacing. The line spacing is adjusted in the eBook to preview the result.

4. Press the ◉ button. The line spacing is updated.

Back to Top

## 10. Changing the Number of Words Per Line

While reading an eBook, you may change the number of words that appear on one line, which changes the size of the left and right margins. To change the number of words per line:

1. Press the ◉ button. The eBook menu appears.

2. Select **Change Font Size** using the ⬜ controller and press the center. The

*Figure 6: Bookmarked Page*

# 6. Viewing Your Notes, Highlights, and Bookmarks

While reading an eBook or periodical, you may view a list of all of your bookmarks, notes, and highlights in order to navigate to each directly. To view a list of your bookmarks, notes, and highlights:

1. Touch anywhere on the screen. The eBook menu appears at the bottom of the screen.
2. Touch the ▤ button. The Content menu appears, as shown in **Figure 7**. A list of your Notes, Highlights, and Bookmarks are shown at the bottom of the screen under 'My Notes & Marks'.
3. Touch an item in the list. The Kindle Fire navigates to its location.

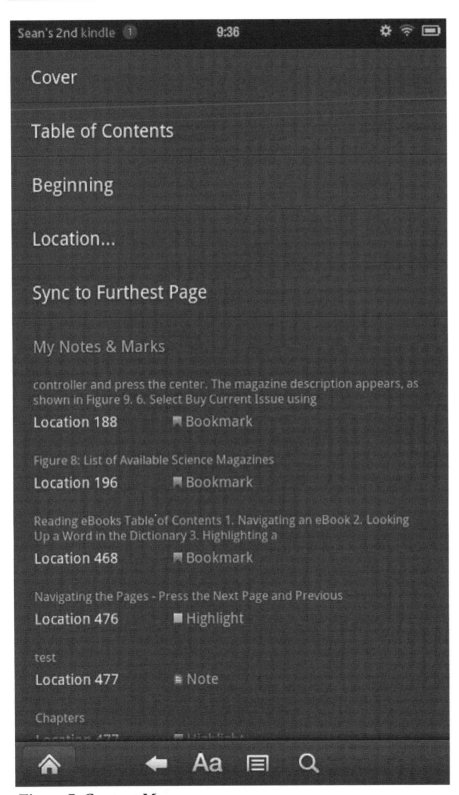

*Figure 7: Content Menu*

# 7. Changing the Text Size

While reading an eBook, the size of the text can be changed. To change the text size:

1. Touch anywhere on the screen (as long as it is not a link). The eBook menu appears at the bottom of the screen.
2. Touch the **Aa** button. The Font menu appears, as shown in **Figure 8**.
3. Touch one of the **Aa** icons. The text size is changed. Touch anywhere outside of the Font menu to return to reading where you left off.

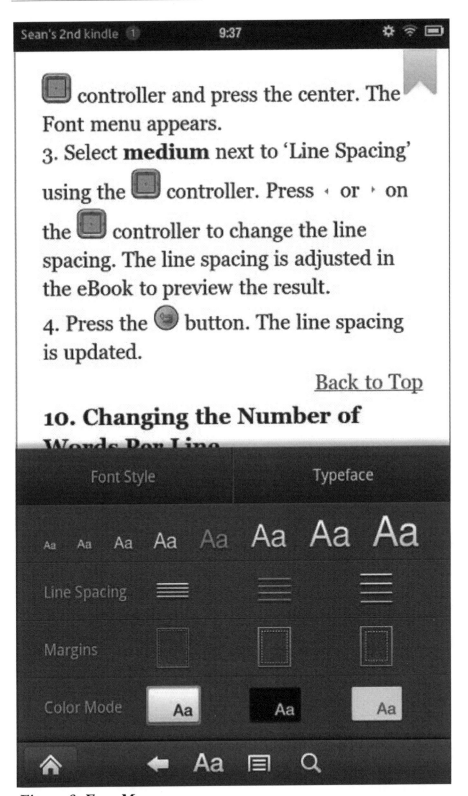

controller and press the center. The Font menu appears.

3. Select **medium** next to 'Line Spacing' using the controller. Press ‹ or › on the controller to change the line spacing. The line spacing is adjusted in the eBook to preview the result.

4. Press the button. The line spacing is updated.

Back to Top

## 10. Changing the Number of Words Per Line

*Figure 8: Font Menu*

# 8. Changing the Font

While reading an eBook, the type of font displayed can be changed. To change the font:

1. Touch anywhere on the screen (as long as it is not a link). The eBook menu appears at the bottom of the screen.
2. Touch the **Aa** button. The Font menu appears.
3. Touch **Typeface**. The Typeface menu appears. Touch the menu and move your finger up or down to scroll through the available fonts.
4. Touch a font. The font is applied. Touch anywhere outside of the Typeface menu to return to reading where you left off.

# 9. Changing the Line Spacing

While reading an eBook, the amount of space between each line can be altered for easier reading. To change the line spacing:

1. Touch anywhere on the screen (as long as it is not a link). The eBook menu appears at the bottom of the screen.
2. Touch the **Aa** button. The Font menu appears.
3. Touch the ▤ icon, ▤ icon, or ▤ icon. The new line spacing is applied. Touch anywhere outside of the Font menu to return to reading where you left off.

# 10. Changing the Page Margins

While reading an eBook, you may change the sizes of the left, right, top, and bottom margins. To change the page margins:

1. Touch anywhere on the screen (as long as it is not a link). The eBook menu appears at the bottom of the screen.
2. Touch the **Aa** button. The Font menu appears.
3. Touch the ▢ icon, ▢ icon, or ▢ icon. The new page margins are applied. Touch anywhere outside of the Font menu to return to reading where you left off.

# 11. Changing the Color Mode

While reading an eBook, you may change the Color mode. The available options are black on white (the default), white on black, and black on sepia. To change the Color mode:

1. Touch anywhere on the screen (as long as it is not a link). The eBook menu appears at the bottom of the screen.
2. Touch the **Aa** button. The Font menu appears.
3. Touch the **Aa** icon, **Aa** icon, or **Aa** (sepia colored) icon. The new Color mode is applied. Touch anywhere outside of the Font menu to return to reading where you left off.

# 12. Searching an eBook

You may search an eBook for a particular word or phrase. To search an eBook:

1. Touch anywhere on the screen (as long as it is not a link). The eBook menu appears at the bottom of the screen.
2. Touch the 🔍 button. The Search field appears at the top of the screen.
3. Type a search word or phrase and touch the **Go** button. A list of available results appears, as shown in **Figure 9**. Touch a search result to navigate to its location.

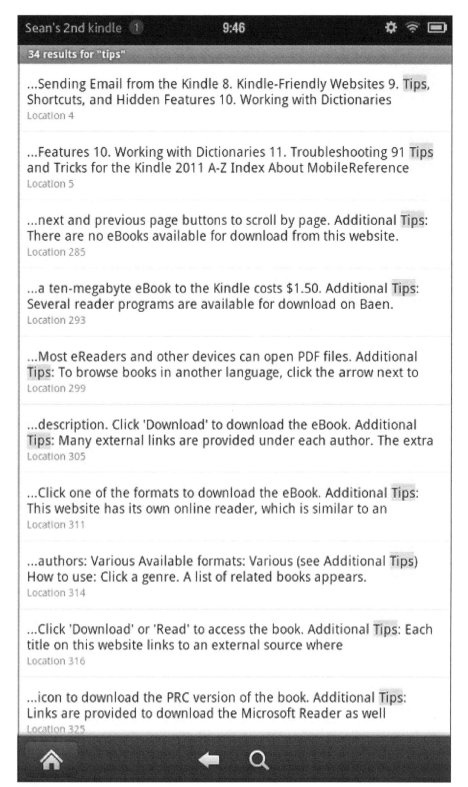

*Figure 9: Search Results*

# Managing Movies and TV Shows

## Table of Contents

## 1. Browsing Movies and TV Shows in the Video Store

You can browse the Amazon library of movies and TV shows right on your Kindle Fire. To browse movies and TV shows:

1. Touch **Video** at the top of the Library. The Video store opens, as shown in **Figure 1**.
2. Touch **View all** next to 'Movies' or 'TV Shows'. A list of popular movies or TV shows appears, as shown in **Figure 2**.
3. Touch a category at the top of the page to browse movies or TV shows. Touch and hold a category and slide your finger to the left or right to view more categories. You can also search for a movie or TV show by touching **Search Video Store** at the top of the screen.

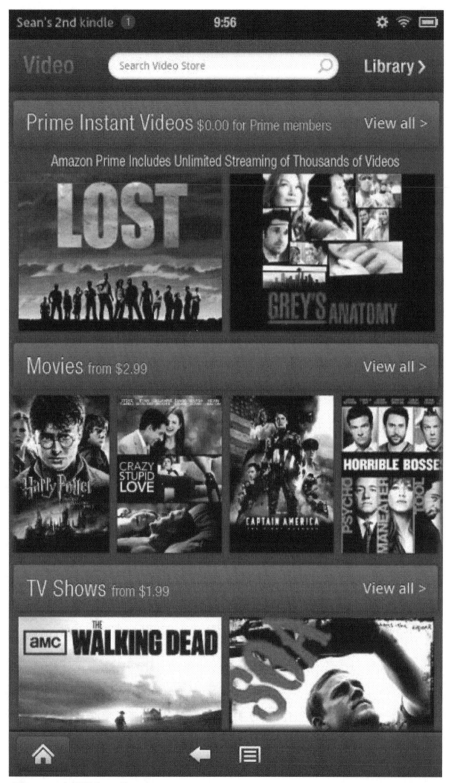

*Figure 1: Video Store*

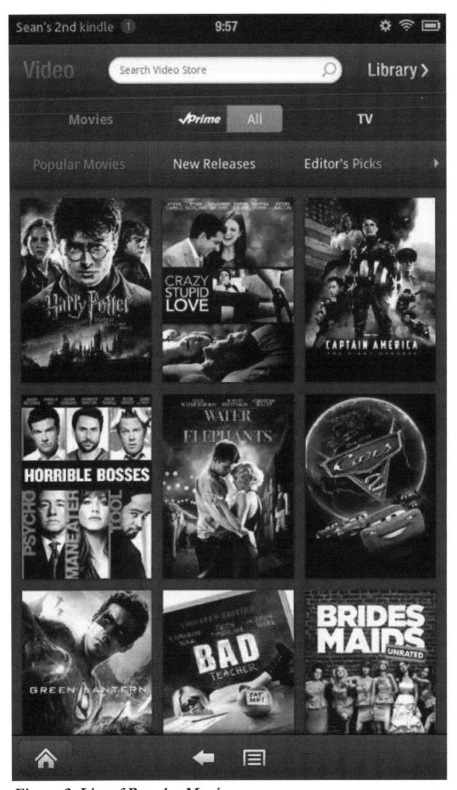

*Figure 2: List of Popular Movies*

# 2. Buying or Renting a Movie

You can purchase or rent movies from the Amazon library using your Kindle Fire. To buy or rent a movie:

***Warning: There is no confirmation after touching the*** **`Rent`** ***button when renting a movie. Make sure you want the rental, as there are no refunds.***

1. Find the movie you wish to buy or rent. Refer to *"Browsing Movies and TV Shows in the Video Store"* on page 71 in the Video Store to learn how.
2. Touch the movie thumbnail. The Movie description appears, as shown in **Figure 3**.
3. Touch the price next to one of the rental or purchase options or touch the **`See all`** button to see more options. A Movie Purchase confirmation appears, as shown in **Figure 4**.
4. Touch the **`Confirm`** button. The movie is purchased and downloaded to your Kindle Fire. Touch **Download** to load the movie onto your Kindle Fire or **Watch Now** to stream it using a Wi-Fi connection.

*Note: When renting a movie, you have 30 days to begin watching it, with the movie expiring 48 hours after you touch 'Watch Now' or 'Download'. Downloading a movie to the Kindle Fire allows you to watch it while not connected to the internet. However, you may only download the movie to one device at a time and cannot stream it on another device registered to your Amazon account while it is loaded onto the Kindle Fire.*

*Figure 3: Movie Description*

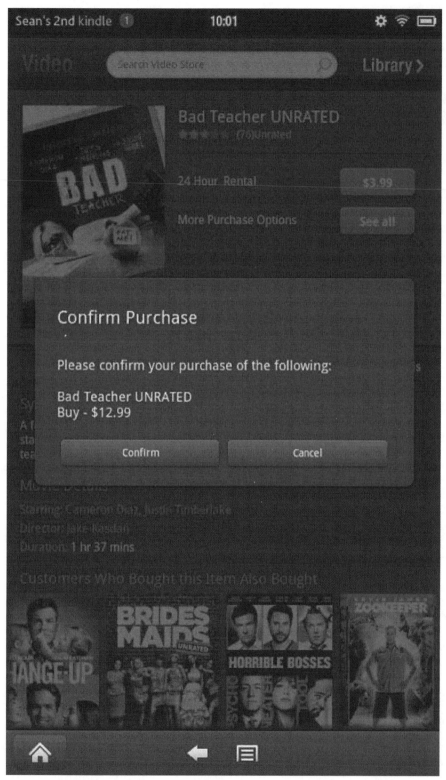

*Figure 4: Movie Purchase Confirmation*

# 3. Buying or Renting a TV Show

You can purchase or rent TV shows from the Amazon library using your Kindle Fire. To buy or rent a TV show:

***Warning: There is no confirmation after touching the*** **Buy** ***button when purchasing or renting a TV show. Make sure you that want the show, as there are no refunds.***

1. Find the TV show you wish to buy or rent. Refer to *"Browsing Movies and TV Shows in the Video Store"* on page 71 to learn how.
2. Touch the movie thumbnail. The TV Show description and a list of episodes appear, as shown in **Figure 5**.
3. Touch the price next to one of the rental or purchase options or touch the **See all** button to see more options. The **Buy** button appears.
4. Touch the **Buy** button. The TV show is purchased. Touch **Download** to load the TV show onto your Kindle Fire or **Watch Now** to stream it using a Wi-Fi connection.

*Note: Downloading a TV show to the Kindle Fire allows you to watch it while not connected to the internet. However, you may only download the TV show to one device at a time and cannot watch it on another device registered to your Amazon account while it is loaded onto the Kindle Fire. When renting a TV show, you have 30 days to begin watching it before it expires.*

*Figure 5: TV Show Description*

# 4. Playing a Movie or TV Show

The Kindle Fire can play movies and TV shows from the library. To play a movie or TV show:

1. Touch **Video** at the top of the Library. The Video store opens.
2. Touch **Library** at the top right of the screen. The Video library opens, as shown in **Figure 6**.
3. Touch **Movies** or **TV**. The corresponding library opens.
4. Touch a video thumbnail. The movie description appears.
5. Touch the [Watch Now] button. The video begins to play.
6. Use the following tips to control the playback of a video:

- **Controlling the Volume** - Touch the screen anywhere. The Video controls appear, as shown in **Figure 7**. Touch the ⬤ on the ▬○▬ bar and drag it to the left to decrease the volume or to the right to increase it.

- **Pausing and Resuming the Video** - Touch the screen anywhere. The Video controls appear. Touch the ⏸ button. The video is paused. Touch the ▶ button. The video resumes playing.

- **Rewinding by Ten Seconds** - Touch the screen anywhere. The Video controls appear. Touch the [10s] button. The video rewinds by ten seconds and resumes playing. You can also rewind while the movie is paused. When you touch the ▶ button, the movie will resume from the new location.

- **Navigating to a Specific Location** - Touch the screen anywhere. The Video controls appear. Touch the ⬤ on the ▬▬○▬▬ bar and drag it to the desired location. The video skips to the location and continues to play.

*Note: If a video has not downloaded or buffered completely, you will be unable to navigate to a location which has not yet finished loading.*

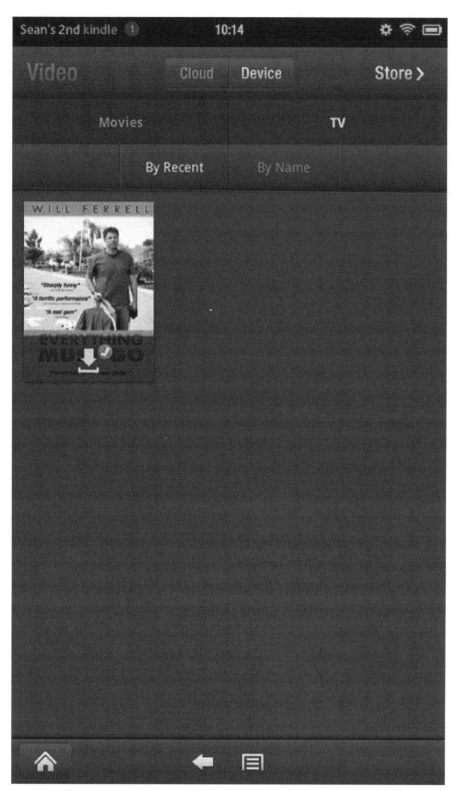

*Figure 6: Video Library*

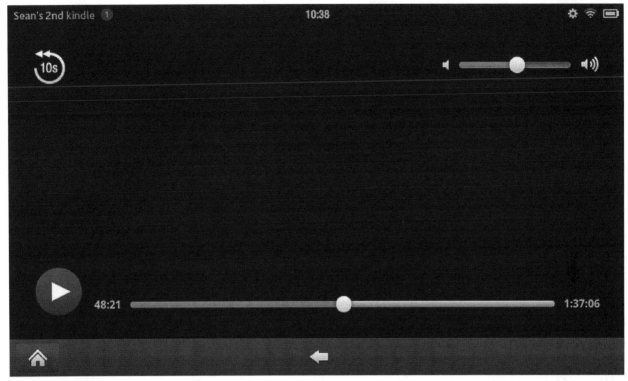

*Figure 7: Video Controls*

# 5. Archiving Movies and TV Shows

Any movies or TV shows that are stored on your Kindle Fire can be archived and stored in the Amazon Cloud where they do not take up space on your device. To archive movies and TV shows:

1. Touch **Video** at the top of the Library. The Video store opens.
2. Touch **Library** at the top right of the screen. The Video library opens.
3. Touch **Movies** or **TV** to find the video you wish to archive. The corresponding video library opens.
4. Touch and hold a video thumbnail. The Video menu appears, as shown in **Figure 8**.
5. Touch **Delete Download**. The video is archived in the Amazon Cloud.

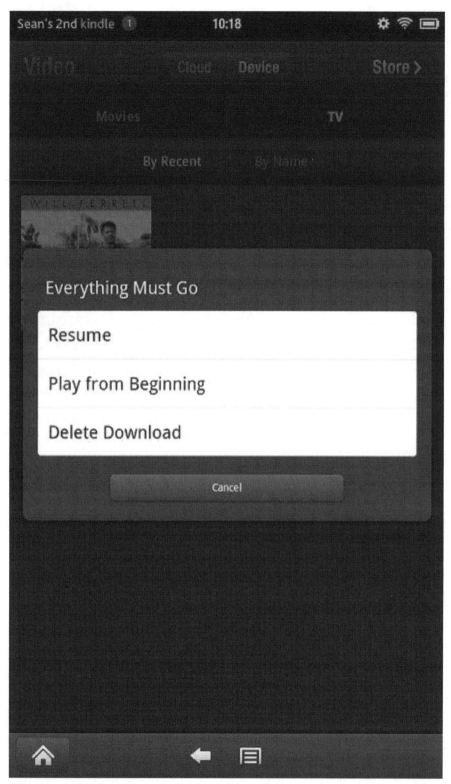

*Figure 8: Video Menu*

# 6. Importing Movies from an Outside Source Using Your PC or Mac

Movies that you have purchased or downloaded elsewhere can be imported to the Kindle Fire. Supported video formats include MP4 and VP8. To import movies:

1. Connect the Kindle Fire to your computer using the provided USB cable. The USB Connected screen appears, as shown in **Figure 9**.
2. Open **My Computer** on a PC and double-click the 'KINDLE' removable drive or double-

   click the  icon on a Mac. The Kindle Folders open on a PC, as shown in **Figure 10**, or on a Mac, as shown in **Figure 11**.
3. Double-click the **Video** folder. The Video folder opens.
4. Drag and drop a video into the Video folder. The video is copied and will appear in the Gallery application on your Kindle Fire, given that it is of the correct format.

*Note: To access the transferred video, touch* **Apps** *at the top of the screen and then touch the* icon.

*Figure 9: USB Connected Screen*

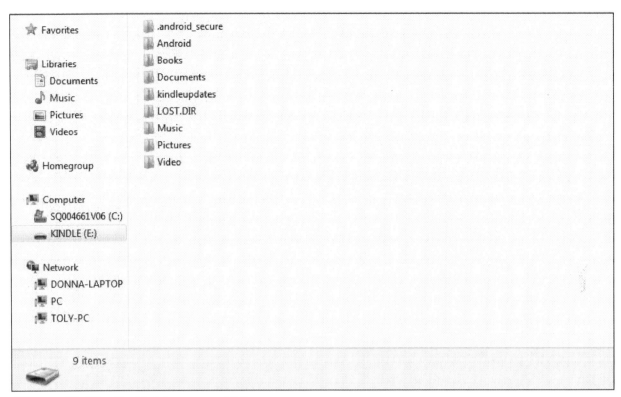

*Figure 10: Kindle Folders on a PC*

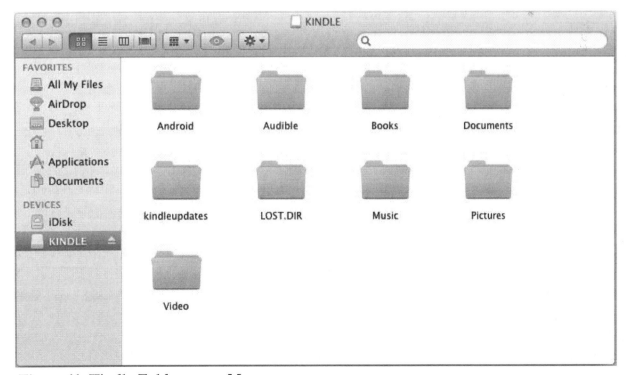

*Figure 11: Kindle Folders on a Mac*

# Managing Music

## Table of Contents

## 1. Browsing Music

You can browse the Amazon music store right on your Kindle Fire. To browse music:

1. Touch **Music** at the top of the Library. The Music library appears, as shown in **Figure 1**.
2. Touch **Store** at the top right of the screen. The Music store opens, as shown in **Figure 2**.
3. Touch one of the categories above the album list. The category opens. You can also search for a particular artist, album, or song by touching **Search Music Store** at the top of the screen.
4. Touch an artist. The available albums appear.
5. Touch an album. The available songs appear. Touch the ▶ button next to a song to preview it.

*Note: Refer to* "Buying a Song or Album" *on page 89 to learn how to purchase a song or album.*

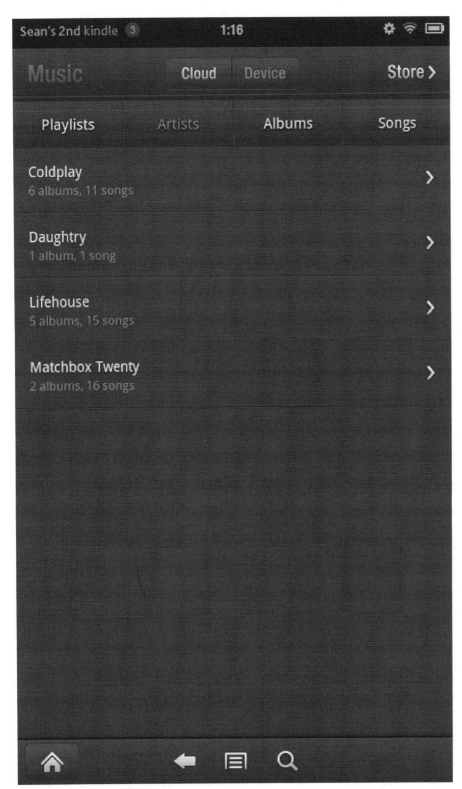

*Figure 1: Music Library*

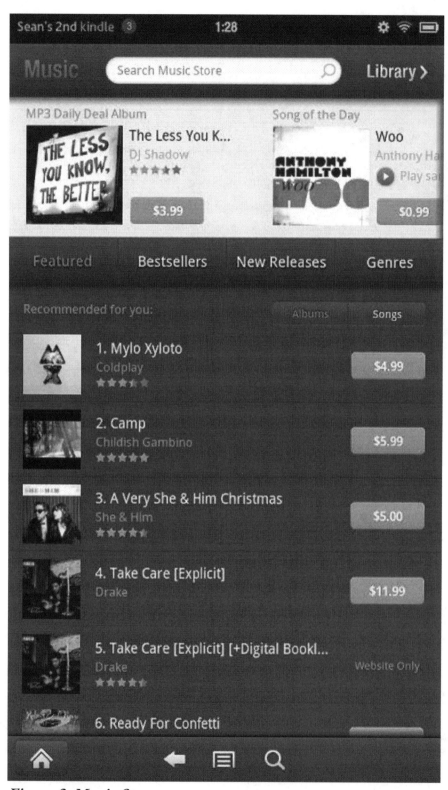

*Figure 2: Music Store*

## 2. Buying a Song or Album

You can purchase music from the Amazon store using your Kindle Fire. To buy a song or album:

1. Find the album containing the song you wish to download. Refer to *"Browsing Music"* on page 86 to learn how.
2. Touch the name of the album. The list of songs appears, as shown in **Figure 3**. Touch the ▶ button next to a song to preview it.
3. Touch the price of a song or album. The [Buy] button appears.
4. Touch the [Buy] button. The song or album is purchased and the Download options appear, as shown in **Figure 4**.
5. Touch the [Save to your Amazon Cloud Drive] button to leave the song in the Amazon Cloud and stream it whenever you wish to listen to it. Choosing this option will not allow you to listen to a song without a Wi-Fi connection. Touch the [Save to this device] button if you wish to be able to listen to a song without a Wi-Fi connection. Choosing this option will take up space on your device, but will also save the music in the Amazon Cloud.

*Note: Refer to* "Tips and Tricks" *on page 184 to learn how to set the Kindle Fire to automatically download a song to the cloud or device when buying music.*

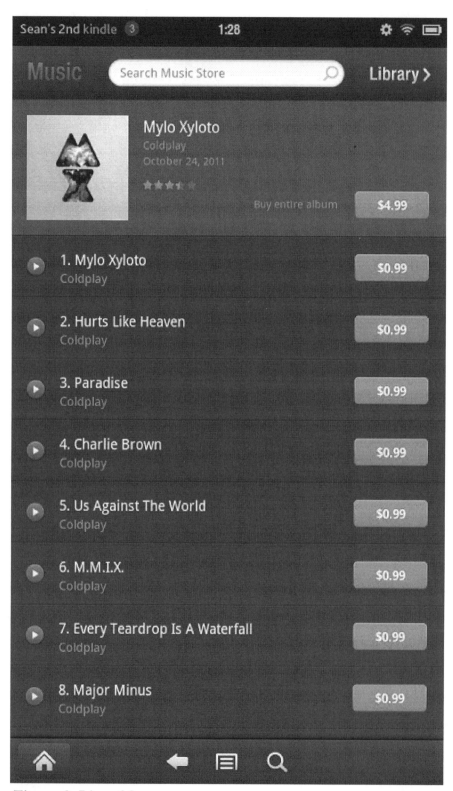

*Figure 3: List of Songs*

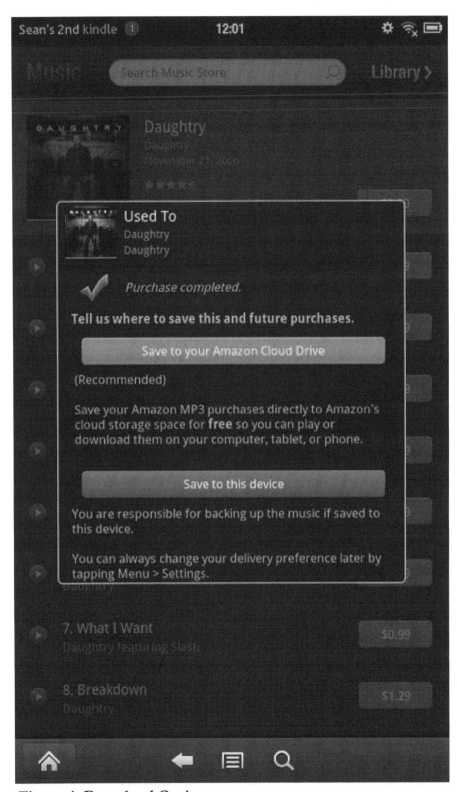

*Figure 4: Download Options*

# 3. Playing a Song

You can play music using your Kindle Fire. To play a song:

*Note: Touch the* *button next to a song in the Cloud to download it to your device.*

1. Touch **Music** at the top of the Library. The Music library appears
2. Touch **Cloud** or **Device**, depending on where your music is stored. The corresponding storage location opens.
3. Touch a category at the top of the music list (**Playlists**, **Artists**, etc.). The corresponding category appears.
4. Touch an artist to view the corresponding albums and touch an album to view the corresponding songs. Touch a song. The song begins to play, as shown in **Figure 5**.

*Note: To add an entire artist or album to the 'Now Playing' list, touch and hold the corresponding item and touch* **Add artist to Now Playing** *or* **Add album to Now Playing**.

*Figure 5: Song Playing*

# 4. Controlling the Music

While a song is playing, touch one of the following buttons to perform the associated action:

**⏸** - Pauses the current song.

**▶** - Resumes the current song when it is paused.

**🔀** - Shuffles the songs in the current playlist.

**🔁** - Repeats the current playlist. If only one song is enqueued, the song will repeat.

**⏮** - Navigates to the beginning of the current song. If already at the beginning of a song, touching this button will navigate to the previous song.

**⏭** - Skips the current song. If more than one song is enqueued, the next song begins to play.

**⚪ on the ▬▬○▬▬ bar at the bottom of the screen** - Adjusts the volume of the music.

**⚪ on the ▬▬○▬▬ bar below the album art** - Navigates to the desired location in the current song.

**☰** - Opens the music library while music is playing. Touch the name of the song at the bottom of the screen to return to the 'Now Playing' screen at any time.

*Note: Refer to* "Tips and Tricks" *on page 184 to learn how to make music controls appear on the Lock screen.*

# 5. Creating and Editing a Playlist

Playlists can be created and edited right on the Kindle Fire.

To create a playlist:

1. Touch **Music** at the top of the Library. The Music library appears
2. Touch **Device**. The music stored on your device appears.
3. Touch **Playlists**. The existing playlists appear, as shown in **Figure 6**.
4. Touch **Create new playlist**. The New Playlist window appears, as shown in **Figure 7**.
5. Type a name for the new playlist and touch the [Save] button. The new playlist is created and the song list appears.
6. Touch the ⊕ button next to a song. The song is added to the playlist.
7. Touch the [Done] button at the top right of the screen. The playlist is saved.

To edit a playlist:

1. Follow steps 1-3 above. The existing playlists appear.
2. Touch a playlist. The songs within the playlist appear.
3. Touch the [Edit] button at the top right of the screen. The Playlist Editing screen appears, as shown in **Figure 8**.
4. Use the following tips to edit a playlist:

   • Touch the ⊖ button to the right of a song to remove it from the playlist.

   • Touch the [Add] button at the top of the screen to add a song to the playlist. A list of songs appears. Touch the ⊕ button next to a song to add it to the playlist

   • Touch the ⊞ icon to the left of a song and drag it up or down to change the order of the songs.

*Note: You can only create a playlist with the music downloaded to your device.*

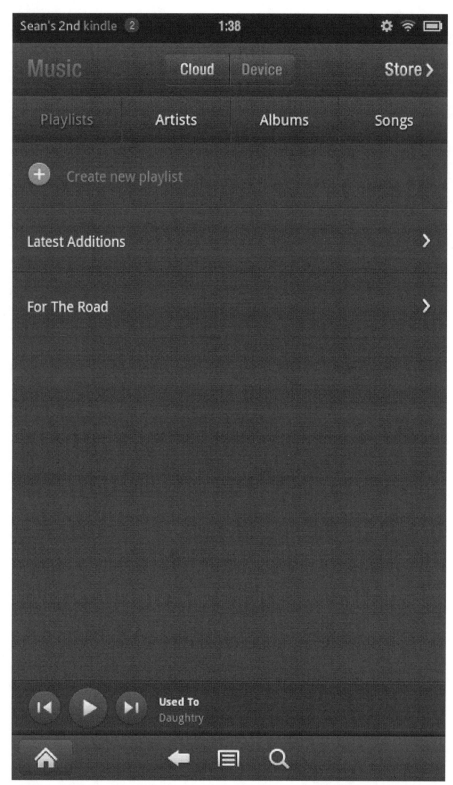

*Figure 6: Existing Playlists*

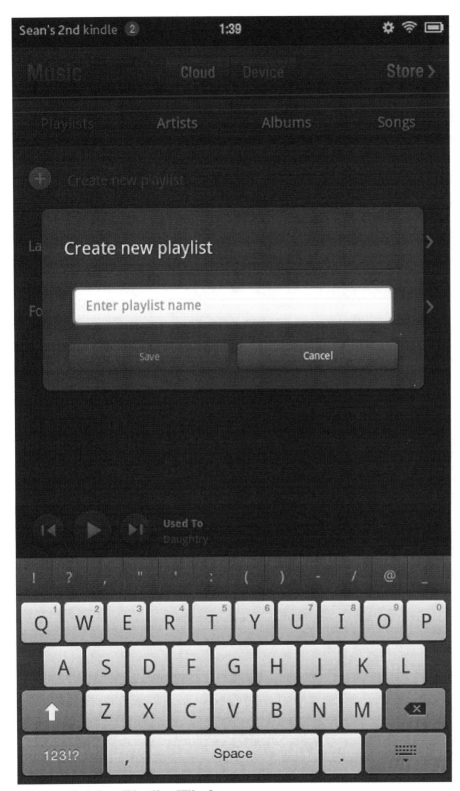

*Figure 7: New Playlist Window*

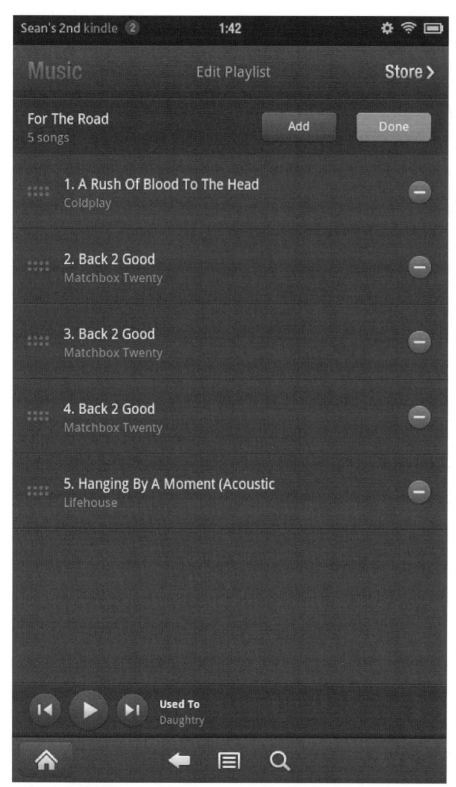

*Figure 8: Playlist Editing Screen*

# 6. Deleting Music

Any music that is stored on your Kindle Fire can be deleted and stored in the Amazon Cloud. To delete music:

*Warning: Any music not backed up in the Amazon Cloud will be permanently deleted from the device and unrecoverable. It is highly recommended to regularly back up all music to your computer. Alternately, you can always touch the* [ Save to your Amazon Cloud Drive ] *button when purchasing music. Refer to* **"Importing Music from an Outside Source Using Your PC or Mac"** *on page 102 to learn how to back up music on your computer. Refer to* **"Buying a Song or Album"** *on page 89 to learn how to purchase music.*

1. Touch **Music** at the top of the Library. The Music library appears
2. Touch **Device**. The music stored on your device appears.
3. Touch and hold an artist, album, or song. The Music menu appears, as shown in **Figure 9**.
4. Touch **Remove all songs by artist from device**, **Remove album from device**, or **Remove song from device**, depending on your selection in step 3. The Music Deletion dialog appears, as shown in **Figure 10**.
5. Touch the [ Yes ] button. The music is removed from your device and, unless it is stored in the Amazon Cloud or backed up on your computer, deleted permanently.

*Note: Any music purchased from the Amazon store is automatically stored in the Amazon cloud when it is removed from the device.*

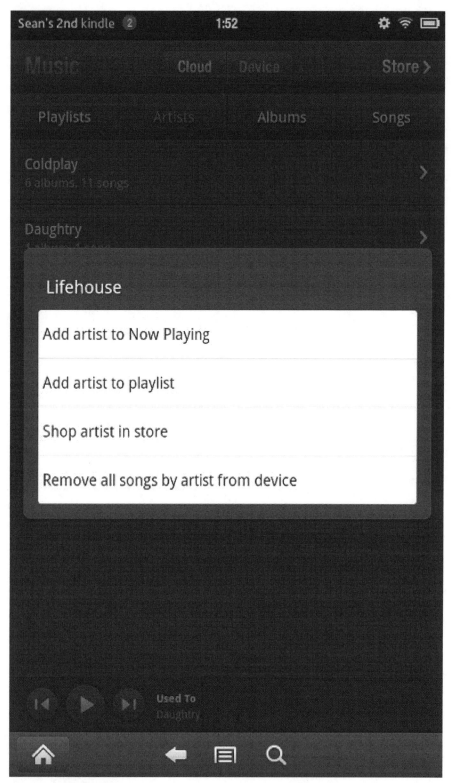

*Figure 9: Music Menu*

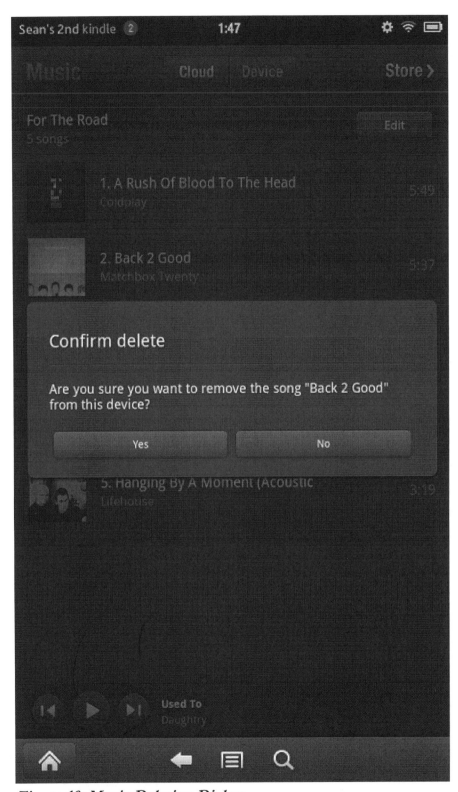

*Figure 10: Music Deletion Dialog*

# 7. Importing Music from an Outside Source Using Your PC or Mac

Music that you have purchased or downloaded elsewhere can be imported to the Kindle Fire. Supported audio formats include non-DRM AAC, MP3, OGG, WAV, and MP4. To import music:

1. Connect the Kindle Fire to your computer using the provided USB cable. The USB Connected screen appears, as shown in **Figure 11**.
2. Open **My Computer** on a PC and double-click the 'KINDLE' removable drive or double-click the ▨▨▨▨ icon on a Mac. The Kindle Folders open on a PC, as shown in **Figure 12**, or on a Mac, as shown in **Figure 13**.
3. Double-click the **Music** folder. The Music folder opens.
4. Drag and drop a song or folder of music into the Video folder. The music is copied and will appear in the Music library.

*Note: You can also transfer any music from the Kindle Fire to your computer by dragging and dropping it from the Music folder to your computer. Even songs purchased in the Amazon Music Store can be transferred. This is recommended to back up your music library.*

*Figure 11: USB Connected Screen*

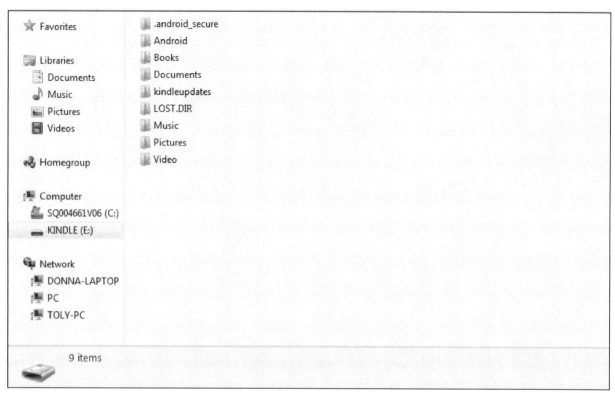

*Figure 12: Kindle Folders on a PC*

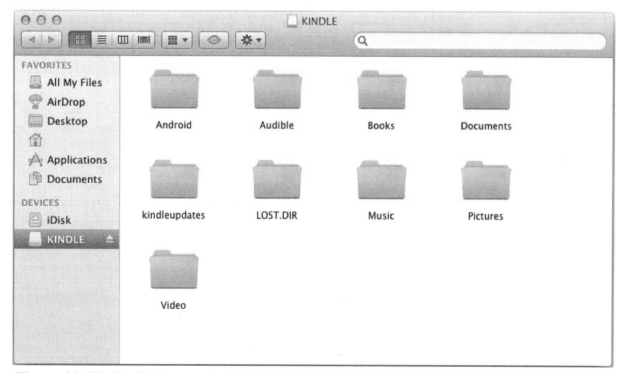

*Figure 13: Kindle Folders on a Mac*

# Using the Email Application

## Table of Contents

## 1. Setting Up the Email Application

Before you can send and receive email using the Email application, you must set it up. To set up the Email application:

1. Touch **Apps** at the top of the Library. The Apps library appears, as shown in **Figure 1**.

2. Touch the 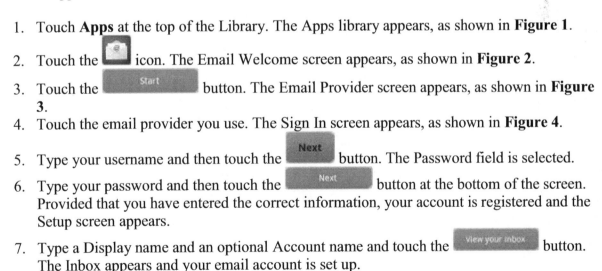 icon. The Email Welcome screen appears, as shown in **Figure 2**.

3. Touch the **Start** button. The Email Provider screen appears, as shown in **Figure 3**.

4. Touch the email provider you use. The Sign In screen appears, as shown in **Figure 4**.

5. Type your username and then touch the **Next** button. The Password field is selected.

6. Type your password and then touch the **Next** button at the bottom of the screen. Provided that you have entered the correct information, your account is registered and the Setup screen appears.

7. Type a Display name and an optional Account name and touch the **View your inbox** button. The Inbox appears and your email account is set up.

*Figure 1: Apps Library*

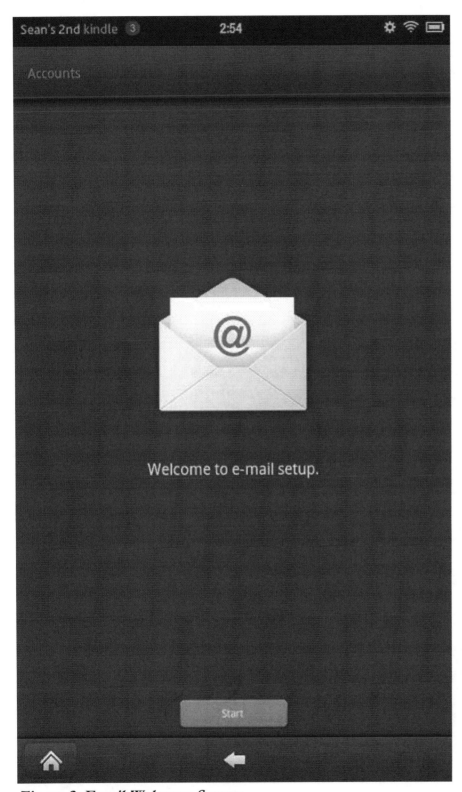

*Figure 2: Email Welcome Screen*

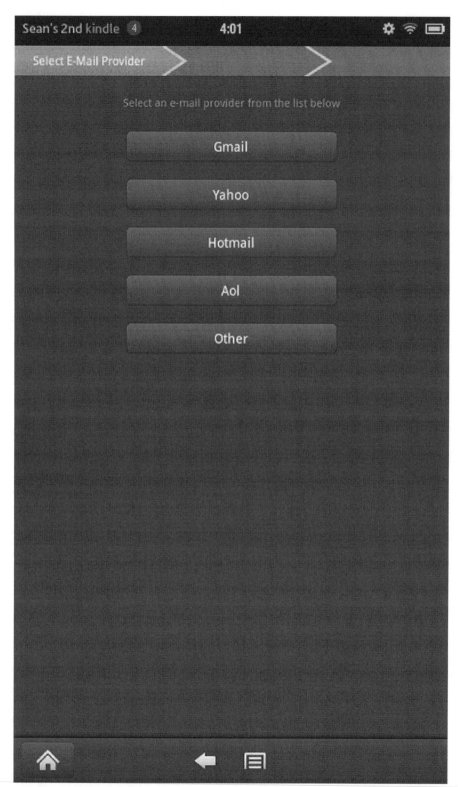

*Figure 3: Email Provider Screen.*

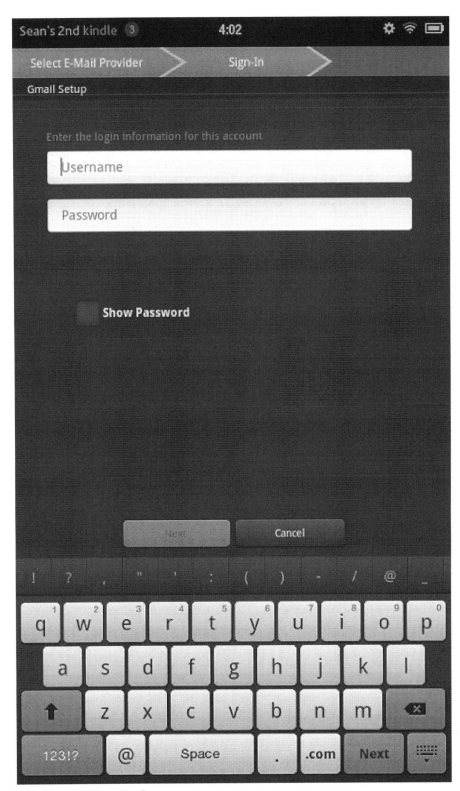

*Figure 4: Sign In Screen*

# 2. Reading Email

You can read email on the Kindle Fire using the Email application. To read email:

1.  Touch **Apps** at the top of the Library. The Apps library appears.

2.  Touch the ◻ icon. A list of registered email accounts appears, as shown in **Figure 5**.
3.  Touch the email account you wish to view. The Email Inbox appears, as shown in **Figure 6**.
4.  Touch an email. The email opens, as shown in **Figure 7**.

*Note: Use the* ◀ *and* ▶ *buttons at the top right of the email to go to the previous or next email, respectively.*

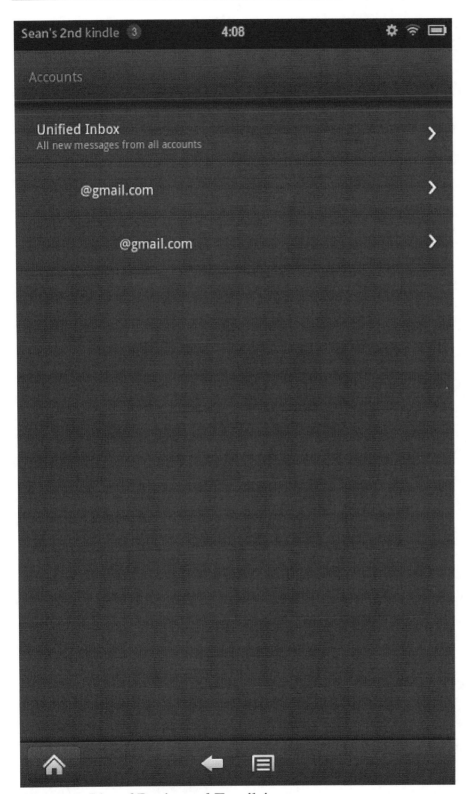

*Figure 5: List of Registered Email Accounts*

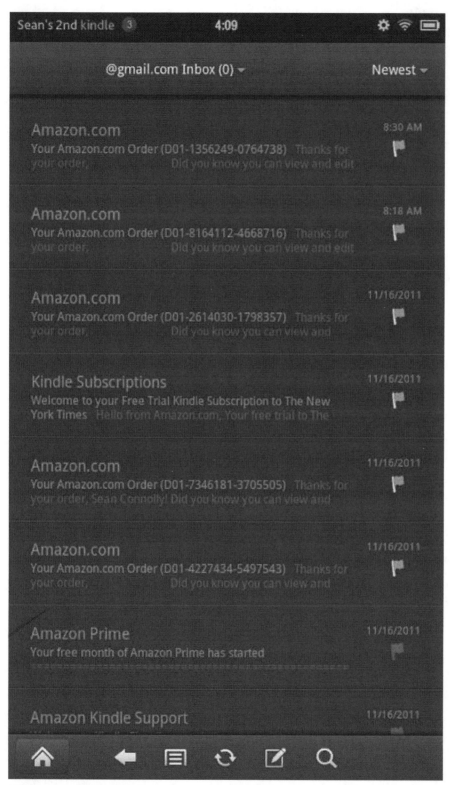

*Figure 6: Email Inbox*

*Figure 7: Open Email*

# 3. Writing an Email

Compose email directly from the Kindle Fire using the Email application. To write an email:

1.  Touch **Apps** at the top of the Library. The Apps library appears.

2.  Touch the [icon] icon. A list of registered email accounts appears.
3.  Touch the email account you wish to view. The Email Inbox appears

4.  Touch the [icon] icon at the bottom of the screen. The New Email screen appears, as shown in **Figure 8**.

5.  Type the recipient's email address or touch the + button to select a stored contact. The address is entered.

6.  Touch the 'Subject' field and type a subject. Touch the 'Message' field and type a message.

7.  Touch the [Send] button. The email is sent.

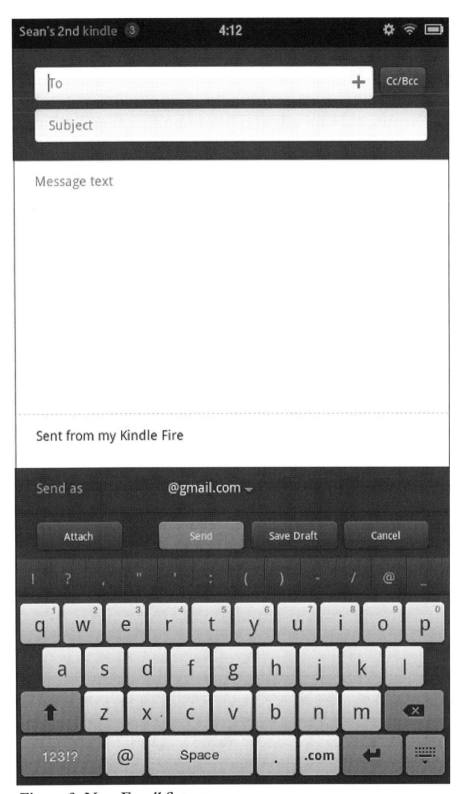

*Figure 8: New Email Screen*

# 4. Replying to and Forwarding Emails

After receiving an email in the Email application, a reply can be sent or the message can be forwarded. To reply to or forward an email:

1. Touch **Apps** at the top of the Library. The Apps library appears.

2. Touch the ▢ icon. A list of registered email accounts appears.

3. Touch the email account you wish to view. The Email Inbox appears.

4. Touch an email. The email opens.

5. Touch the ⬀ icon at the bottom of the screen. The Reply and Forward menu appears at the bottom of the screen, as shown in **Figure 9**.

6. Touch one of the following icons to perform the associated action:

   ⬋ - Creates a new email in reply to the sender of the original email.

   ⬉ - Creates a new email in reply to the sender and all of the recipients of the original email.

   ⬀ - Creates a new email for forwarding, copying the original email and leaving the To field blank.

*Figure 9: Reply and Forward Menu*

# 5. Deleting Emails

Emails can be deleted to free up room in your Inbox. To delete an email:

1.  Touch **Apps** at the top of the Library. The Apps library appears.

2.  Touch the ▢ icon. A list of registered email accounts appears.
3.  Touch the email account you wish to view. The Email Inbox appears.
4.  Touch an email. The email opens.

5.  Touch the 🗑 icon at the bottom of the screen. The email is deleted.

# 6. Moving an Email to a Different Folder

Moving emails between folders, such as 'Work' or 'Personal', can be a helpful organizational tool. To move an email to a different folder:

1.  Touch **Apps** at the top of the Library. The Apps library appears.

2.  Touch the ▢ icon. A list of registered email accounts appears.
3.  Touch the email account you wish to view. The Email Inbox appears.
4.  Touch an email. The email opens.
5.  Touch the ▤ button at the bottom of the screen. The Email menu appears at the bottom of the screen, as shown in **Figure 10**.
6.  Touch the ▣ icon. A list of email folders appears, as shown in **Figure 11**.
7.  Touch a folder in the list. The email is moved to the selected folder.

*Figure 10: Email Menu*

*Figure 11: List of Email Folders*

# 7. Searching the Inbox

To find a message in the Inbox, use the search function, which searches email addresses, message text, and subject lines. To search the Inbox:

1. Touch **Apps** at the top of the Library. The Apps library appears.

2. Touch the [icon] icon. A list of registered email accounts appears.

3. Touch the email account you wish to view. The Email Inbox appears.

4. Touch the [icon] button. The Email Search field appears at the top of the screen, as shown in **Figure 12**.

5. Type a search word or phrase and touch the [Search] button on the keyboard. A list of matching emails appears.

*Note: Refer to* "Reading Email" *on page 110 to learn how to view an email.*

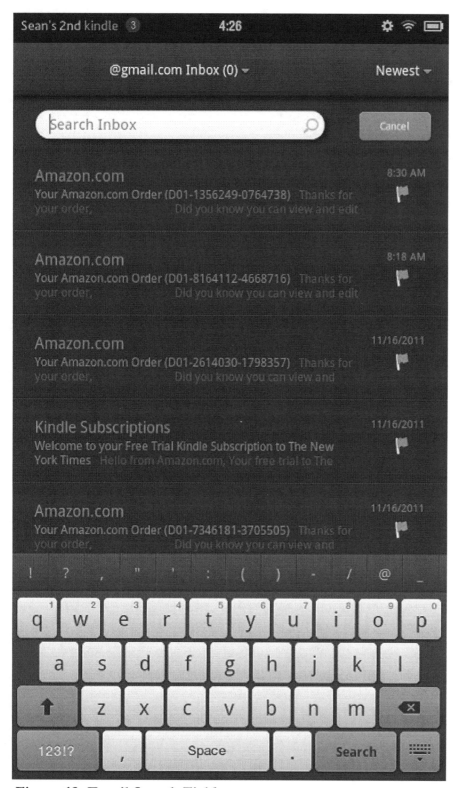

*Figure 12: Email Search Field*

# 8. Changing the Incoming Email Notification Sound

The sound that plays when a new email arrives can be customized for each email account registered to your Kindle Fire. To change the incoming email notification sound:

1. Touch **Apps** at the top of the Library. The Apps library appears.

2. Touch the ▣ icon. A list of registered email accounts appears.
3. Touch the email account you wish to edit. The Inbox appears.
4. Touch the ▤ button at the bottom of the screen. The Email menu appears.

5. Touch the ⚙ icon. The Settings screen appears, as shown in **Figure 13**.
6. Touch the screen and move your finger up to scroll down in the list. Touch Notification Sound. A list of Notification Sounds appears, as shown in **Figure 14**.

*Note: You must have your Email application set to "Push" in order to be notified of new emails when they arrive. Touch* **Fetch new messages** *on the Settings screen and then touch* **Push** *to turn the feature on.*

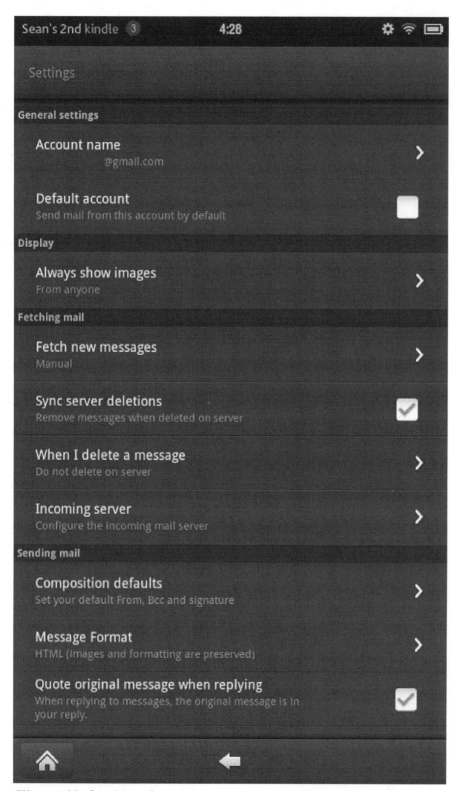

*Figure 13: Settings Screen*

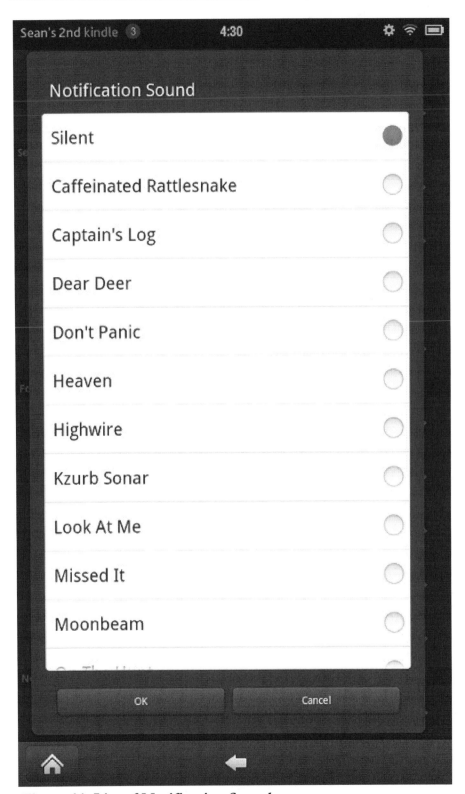

*Figure 14: List of Notification Sounds*

# Managing Contacts

## Table of Contents

## 1. Adding a New Contact

The Kindle Fire allows you to manage your contacts using the Contacts application. To add a new contact:

1. Touch **Apps** at the top of the Library. The Apps screen appears, as shown in **Figure 1**.

2. Touch the ![icon] icon. The Contacts application opens, as shown in **Figure 2**.

3. Touch the ![button] button at the bottom of the screen. The Contacts menu appears, as shown in **Figure 3**.

4. Touch the ![button] button at the bottom of the screen. The New Contact screen appears, as shown in **Figure 4**.

5. Touch each field and type the required information. Touch the ![Save Changes] button at the bottom of the screen. The contact is added.

*Figure 1: Apps Screen*

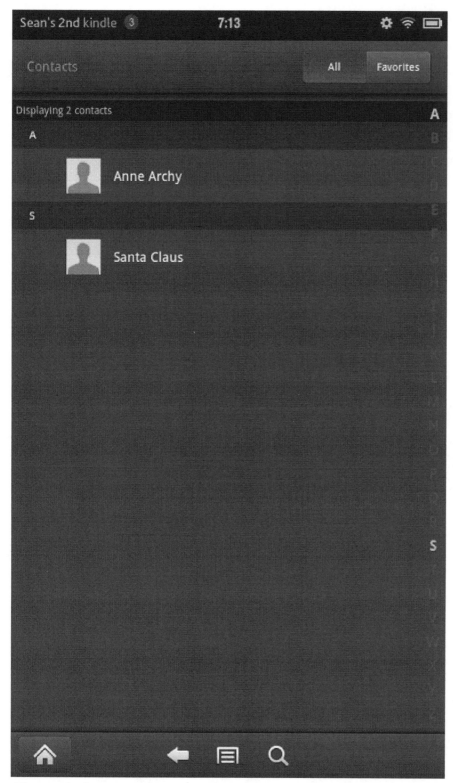

*Figure 2: Contacts Application*

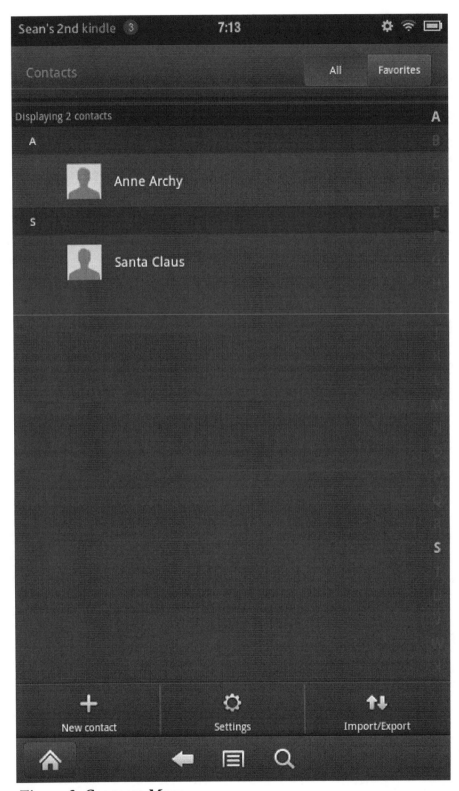

*Figure 3: Contacts Menu*

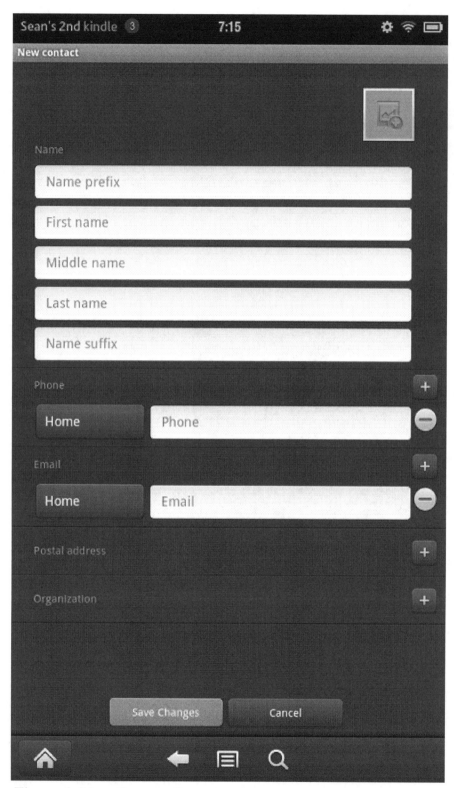

*Figure 4: New Contact Screen*

# 2. Editing a Contact's Information

Edit a contact's information to add additional data, such as an email address or additional phone number. To edit an existing contact's information:

1. Touch **Apps** at the top of the Library. The Apps screen appears.

2. Touch the ![icon] icon. The Contacts application opens.
3. Touch and hold a contact's name. The Contact Info menu appears, as shown in **Figure 5**.
4. Touch **Edit contact**. The contact's information appears.

5. Touch each field to edit the contact's information. Touch the [ Save Changes ] button at the bottom of the screen. The contact's information is saved.

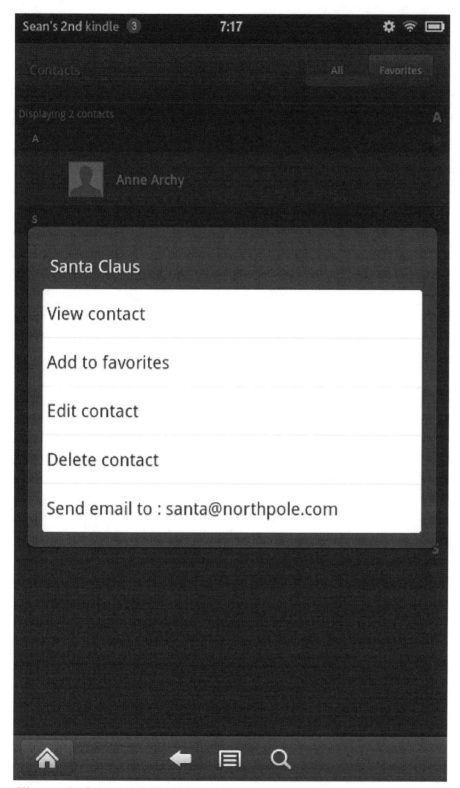

*Figure 5: Contact Info Menu*

# 3. Deleting a Contact

Free up memory by deleting contacts from the Kindle Fire. To delete an unwanted contact:

1. Touch **Apps** at the top of the Library. The Apps screen appears.

2. Touch the ![icon] icon. The Contacts application opens.
3. Touch and hold a contact's name. The Contact Info menu appears.
4. Touch **Delete contact**. A confirmation dialog appears.

5. Touch the ![OK] button. The contact is deleted.

# 4. Sharing a Contact's Information

You can share a contact's information with others via email. To share a contact's information:

1. Touch **Apps** at the top of the Library. The Apps screen appears.

2. Touch the ![icon] icon. The Contacts application opens.
3. Touch a contact's name. The Contact Info appears.
4. Touch the ![button] button at the bottom of the screen. The Contact Editing menu appears, as shown in **Figure 6**.

5. Touch the ![icon] icon at the bottom of the screen. The New Email screen appears with the contact's information attached, as shown in **Figure 7**.

6. Type the recipient's email address or touch the ➕ button to select a stored contact. The address is entered.
7. Touch the 'Subject' field and type a subject. Touch the 'Message' field and type a message.

8. Touch the ![Send] button. The contact's information is shared.

*Figure 6: Contact Editing Menu*

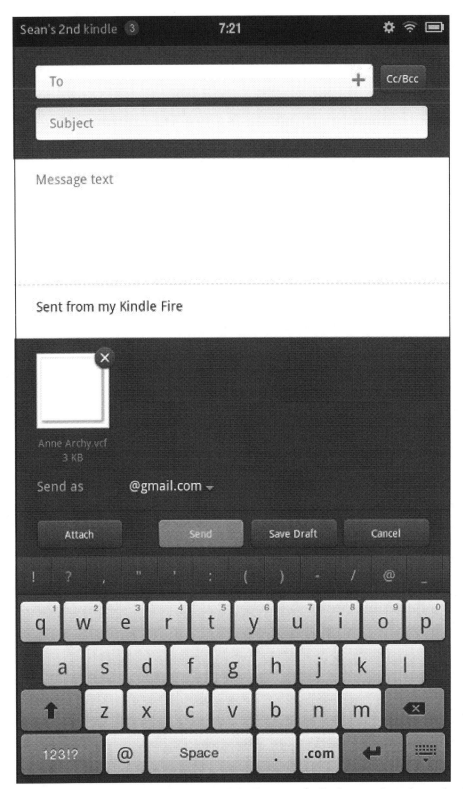

*Figure 7: New Email Screen with Contact's Information Attached*

# Using the Silk Web Browser

## Table of Contents

## 1. Navigating to a Web Page

The Kindle Fire has a built-in Web browser called Silk. To navigate to a web page using its web address:

1.  Touch **Web** at the top right of the Library. The Silk browser opens, as shown in **Figure 1**.
2.  Touch the address field at the top of the screen. The virtual keyboard appears.

3.  Type a web address and touch the ⬚Go⬚ button. The Silk browser navigates to the web page.

*Figure 1: Silk Browser*

# 2. Adding Bookmarks

Web pages can be stored as bookmarks for faster access. To add a web page to your bookmarks:

1.  Navigate to a web page. Refer to *"Navigating to a Web Page"* on page 136 to learn how.

2.  Touch the ▯ icon at the bottom of the screen. The Bookmarks screen appears, as shown in **Figure 2**.

3.  Touch the leftmost thumbnail in the top row, which has a ➕ icon on it. The Add Bookmark dialog appears, as shown in **Figure 3**.

4.  Touch the 'Name' field and type a custom name for the bookmark, if desired. Touch the OK button. The web page is saved to bookmarks and appears as the last item in the last row.

*Note: Touch the ▯ icon at any time to view your bookmarks. Refer to* "Tips and Tricks" *on page 184 to learn how to delete a bookmark*

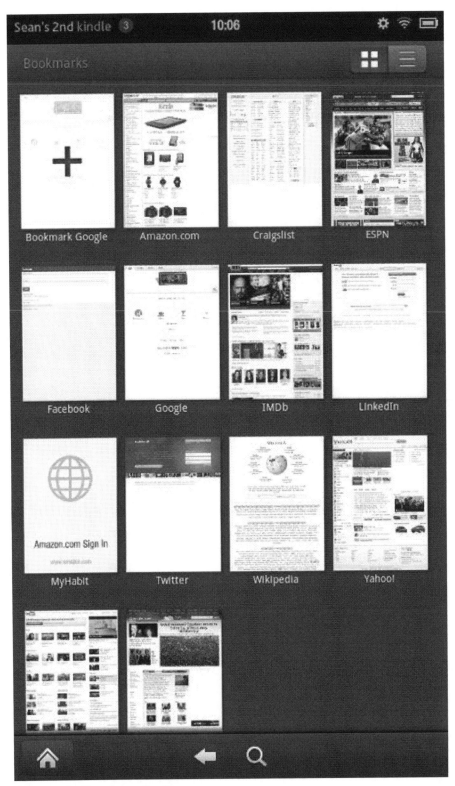

*Figure 2: Bookmarks Screen*

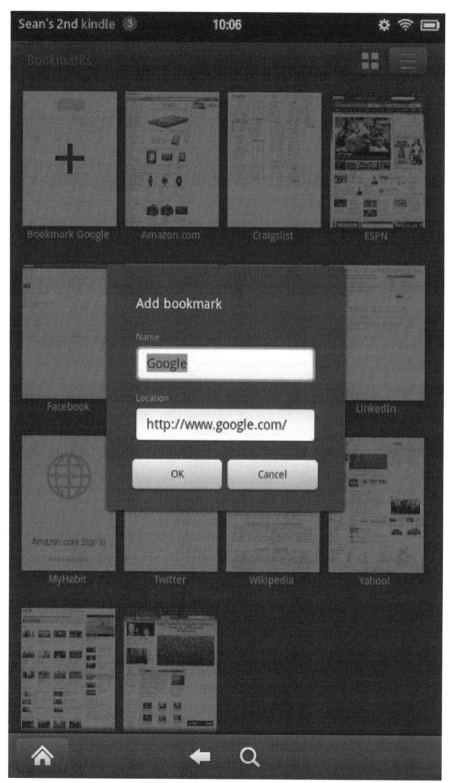

*Figure 3: Add Bookmark Dialog*

# 3. Managing Tabs

Up to ten tabs can be opened at once and displayed at the top of the Silk browser. Use the following tips to manage tabs:

- Touch the ➕ icon at the top right of the Silk browser to open a new tab.

- Touch the ✖ icon on a tab to close it.

- Touch a tab and drag your finger to the left or right to view more open tabs when more than three tabs are open.

*Note: Refer to* "Tips and Tricks" *on page 184 to learn how to close all tabs at once.*

# 4. Using Links

In addition to touching a link to navigate to its destination, there are other link options. Touch and hold a link to see all of the link options, as follows:

- **Open in new tab** - Opens the link in a new tab without closing the existing one.

- **Bookmark Link** - Adds the link to the bookmarks.

- **Copy link URL** - Copies the link to the clipboard, allowing it to be pasted in any text field.

- **Share Link** - Copies the link into an email or a friend's Facebook wall, depending on your selection.

# 5. Searching a Web Page for a Word or Phrase

While surfing the web, any web page can be searched for a word or phrase. To search a web page:

1. Touch the ▤ button at the bottom of the screen. The Browser menu appears, as shown in **Figure 4**.
2. Touch the 🔍 button. The Search field appears at the top of the screen.
3. Type a search word or phrase. The search results are highlighted in orange as you type. Touch the ▲ and ▼ buttons to the left of the search field to scroll through the results on the web page.

*Figure 4: Browser Menu*

# 6. Viewing Recently Visited Websites

The Kindle Fire stores all recently visited websites in its Browsing History. To view the History, touch the ⊟ button at the bottom of the screen and then touch the 🕐 icon. The Browsing History screen appears, as shown in **Figure 5**.

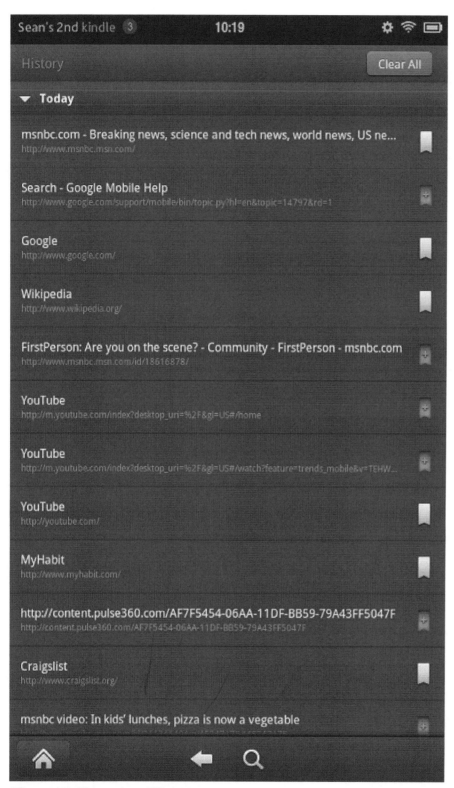

*Figure 5: Browsing History*

# 7. Blocking Pop-Up Windows

Some websites may have pop-up windows that will interfere with web browsing. To block pop-ups:

1. Touch the ▦ button at the bottom of the screen. The Browser menu appears.
2. Touch the ✂ icon. The Browser Settings screen appears, as shown in **Figure 6**.
3. Touch the screen and move your finger up to scroll down. Touch **Block pop-up windows**. The Block Pop-Up Windows menu appears.
4. Touch **Always**. Pop-up windows will now be blocked.

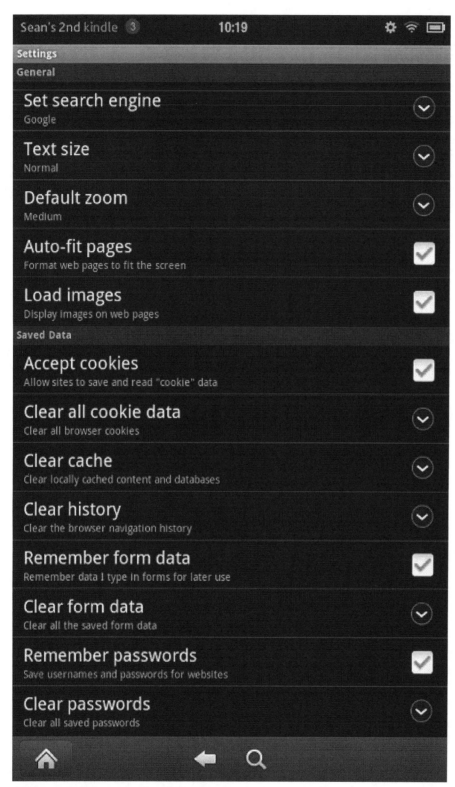

*Figure 6: Browser Settings Screen*

# 8. Changing the Browser Text Size

The size of the text in the Silk Browser can be customized. To change the text size:

1. Touch the ▤ button at the bottom of the screen. The Browser menu appears.
2. Touch the ✖ icon. The Browser Settings screen appears.
3. Touch **Text size**. The Text Size options appear, as shown in **Figure 7**.
4. Touch the preferred text size. The new text size is applied to all text in the Silk browser.

   Touch the ◀ button at the bottom of the screen to resume browsing where you left off.

*Figure 7: Text Size Options*

# 9. Changing the Default Search Engine

When you type a search term in the web address field, the default search engine is used to perform the search. The default search engine is Google. To change the default search engine:

1. Touch the ▤ button at the bottom of the screen. The Browser menu appears.
2. Touch the ⚒ icon. The Browser Settings screen appears.
3. Touch **Set search engine**. A list of available search engines appears, as shown in   **Figure 8**.
4. Touch the preferred search engine. The new default search engine is set and will be used for all searches.

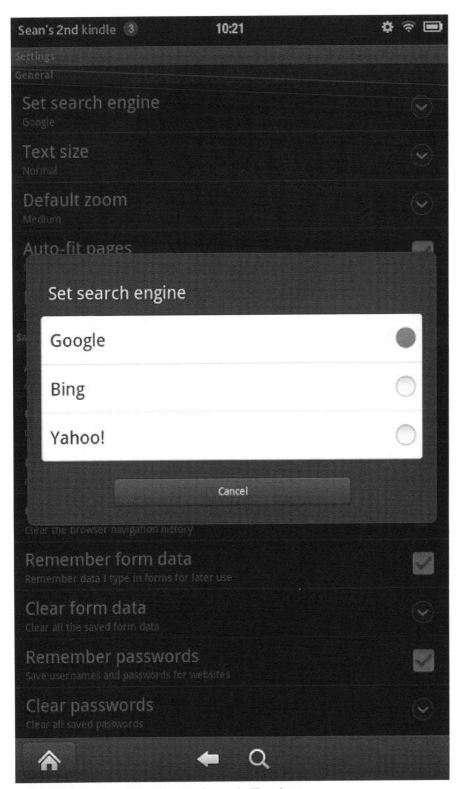

*Figure 8: List of Available Search Engines*

# 10. Saving Passwords in Forms

The Silk browser can automatically save the passwords that you enter in forms. To save passwords in forms:

1. Touch the ▤ button at the bottom of the screen. The Browser menu appears.

2. Touch the ✖ icon. The Browser Settings screen appears.

3. Touch **Remember passwords**. The ✔ appears to the right of 'Remember passwords' and the feature is turned on.

# Managing Applications

## Table of Contents

## 1. Searching for an Application in the Amazon App Store

There are two ways to search for applications on the Kindle Fire:

- **Manual Search** - Search by a specified word
- **Browse by Category** - Games, Travel, Productivity, etc.

**Manual Search**

To search for an application manually:

1. Touch **Apps** at the top of the library. The Apps Library appears, as shown in **Figure 1**.
2. Touch **Store** at the top right of the screen. The Apps Store opens, as shown in **Figure 2**.
3. Touch **Search in Appstore** at the top of the screen. The virtual keyboard appears.
4. Type the name of an application and touch the key. All available application results appear, as shown in **Figure 3**.

**Browse by Category**

To browse applications by category:

1. Touch **Apps** at the top of the library. The Apps Library appears.
2. Touch **Store** at the top right of the screen. The Apps Store opens.
3. Touch a category, such as **Top**, **New**, or **Games**. All of the applications available for that category appear.

*Figure 1: Apps Library*

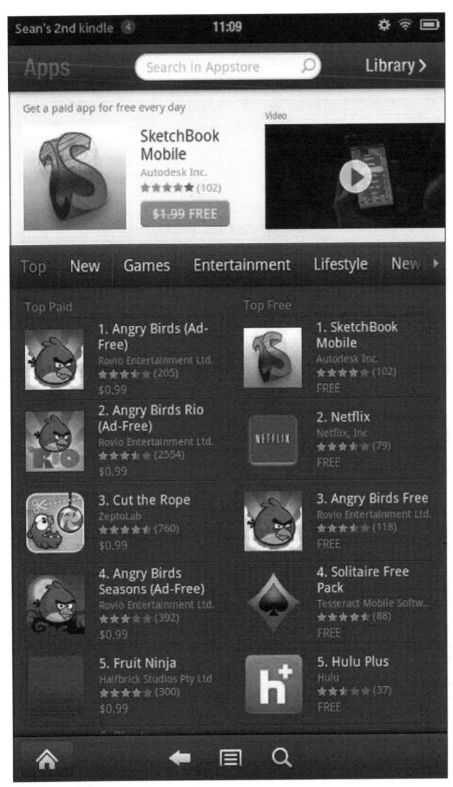

*Figure 2: Apps Store*

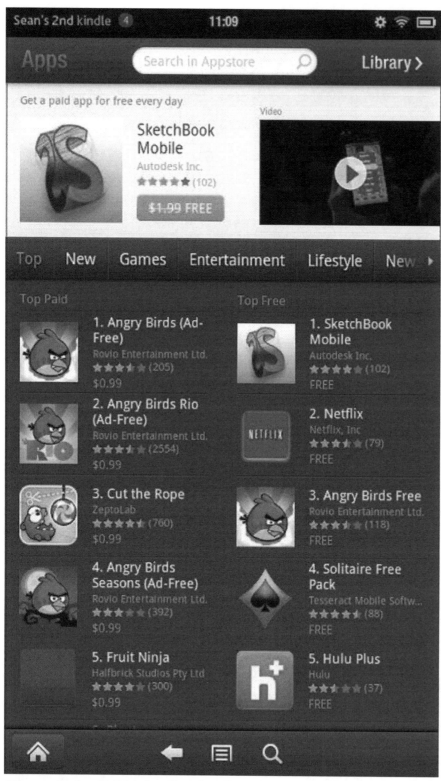

*Figure 3: Available Application Results*

## 2. Buying an Application

Applications can be purchased directly from the Kindle Fire using the Amazon App Store. To buy an application:

***Warning: Before purchasing an application, make sure that you want it. There are no refunds for applications in the Amazon App Store.***

1. Find the application you wish to purchase. Refer to *"Searching for an Application in the Amazon App Store"* on page 153 to learn how. Touch the name of the application. The Application description appears, as shown in **Figure 4**.

2. Touch the price of the application or touch the FREE button. The Get App or Buy App button appears.

3. Touch the Get App or Buy App button. The application is purchased and downloaded to your Apps library.

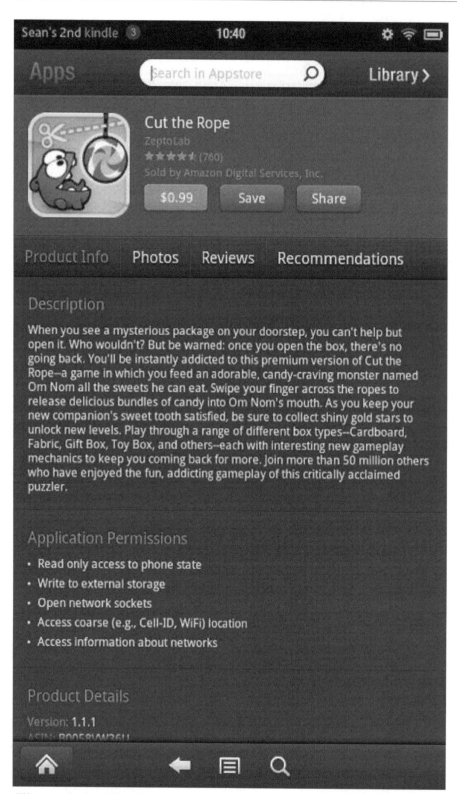

*Figure 4: Application Description*

# 3. Removing an Application from the Kindle Fire

In order to free up memory on the device, you can remove applications from the Kindle Fire and store them in the Amazon cloud. Any application that is removed can always be downloaded later using a Wi-Fi connection. To remove an application:

1. Touch **Apps** at the top of the library. The Apps Library appears.
2. Touch and hold an application icon. The Application options appear above the icon.
3. Touch **Remove from Device**. A confirmation dialog appears.
4. Touch the [ OK ] button. The application is removed from the Kindle Fire.

# 4. Adding an Application to Favorites

Any application can be added to the Favorites, which are located at the bottom of the library, as outlined in **Figure 5**. To add an application to the Favorites:

1. Touch **Apps** at the top of the library. The Apps Library appears.
2. Touch and hold an application icon. The Application options appear above the icon.
3. Touch **Add to Favorites**. The application is added to the Favorites.

*Note: To remove an application from the Favorites, touch and hold it and then touch* **Remove from Favorites**. *The application is removed.*

*Figure 5: Favorites in the Library*

# 5. Reporting an Issue with an Application

If an application malfunctions or is offensive in any way, you can report the issue to Amazon, who will get in touch with the developer to address your concern. To report an issue with an application:

1. Touch **Apps** at the top of the library. The Apps Library appears.
2. Touch **Store** at the top right of the screen. The Apps Store opens.
3. Find the application you wish to report. Refer to *"Searching for an Application in the Amazon App Store"* on page 153 to learn how.
4. Touch the title of the application. The Application description appears.
5. Touch the screen and move your finger up to scroll down to the bottom of the description. Touch **Report an Issue with this App**. The Issue Type menu appears.
6. Touch the type of issue you are having. The appropriate reporting screen appears. Follow the instructions on the screen, which vary depending on your issue.

# 6. Reading User Reviews

Reading user reviews may help when making a decision between similar applications from different developers. To read user reviews for an application:

1. Touch **Apps** at the top of the library. The Apps Library appears.
2. Touch **Store** at the top right of the screen. The Apps Store opens.
3. Find an application. Refer to *"Searching for an Application in the Amazon App Store"* on page 153 to learn how.
4. Touch the title of the application. The Application description appears.
5. Touch **Reviews** below the application icon. The user reviews appear.

# Adjusting the Settings

## Table of Contents

## 1. Adjusting the Brightness

The brightness of the screen can be changed. To adjust the brightness:

1. Touch the ⬛ icon at the top right of the screen. The Quick Settings Banner appears at the top of the screen, as shown in **Figure 1**.

2. Touch the ⬛ icon. The Brightness bar appears.

3. Touch the ⬤ on the ▬▬◯▬▬ bar and drag it to the left to decrease the brightness or to the right to increase it. The brightness is adjusted.

*Figure 1: Quick Settings Banner*

# 2. Setting the Screen Timeout

To avoid unintentional button pushing and save battery life, the Kindle Fire has a feature that allows it to lock itself when it is not in use. The Screen Timeout is the amount of time the Kindle Fire waits before automatically locking itself. To set the Screen Timeout:

1. Touch the ⚙ icon at the top right of the screen. The Quick Settings Banner appears.
2. Touch the ⊕ icon. The Settings screen appears, as shown in **Figure 2**.
3. Touch **Display**. The Display Settings screen appears, as shown in **Figure 3**.
4. Touch **Screen Timeout**. The Screen Timeout menu appears, as shown in **Figure 4**.
5. Touch the preferred Screen Timeout. The new Screen Timeout is set.

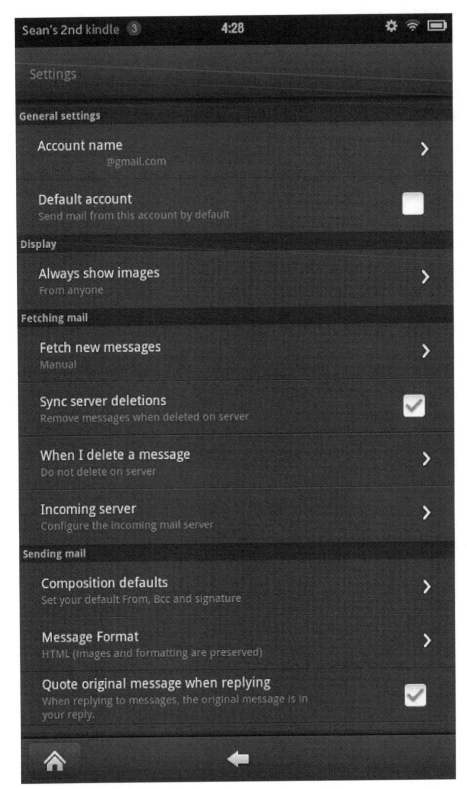

*Figure 2: Settings Screen*

*Figure 3: Display Settings Screen*

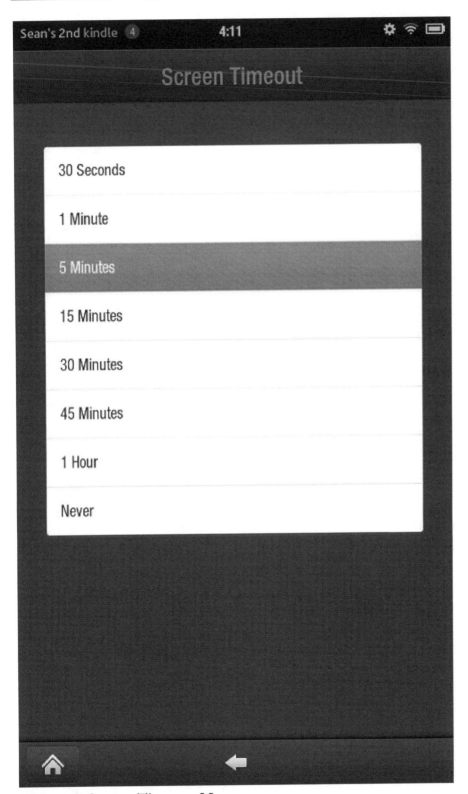

*Figure 4: Screen Timeout Menu*

# 3. Adjusting the Volume

The two speakers on the Kindle Fire are located on top of the device. There are no volume controls on the device itself. To adjust the volume of the speakers:

1. Touch the ⚙ icon at the top right of the screen. The Quick Settings Banner appears.
2. Touch the 🔊 icon. The Volume bar appears.
3. Touch the ⬤ on the ━━━◯━━━ bar and drag it to the left to decrease the volume or to the right to increase it. The volume is adjusted.

# 4. Setting the Notification Sound

The notification sound, such as the one used every time an email arrives or an application sends a notification, can be customized. To set the notification sound:

1. Touch the ⚙ icon at the top right of the screen. The Quick Settings Banner appears.
2. Touch the ⊕ icon. The Settings screen appears.
3. Touch **Sounds**. The Sound Settings screen appears, as shown in **Figure 5**.
4. Touch **Notification Sounds**. The Notification Sound menu appears, as shown in **Figure 6**.
5. Touch the preferred notification sound. The notification sound is set and a preview of the sound plays.

*Figure 5: Sound Settings Screen*

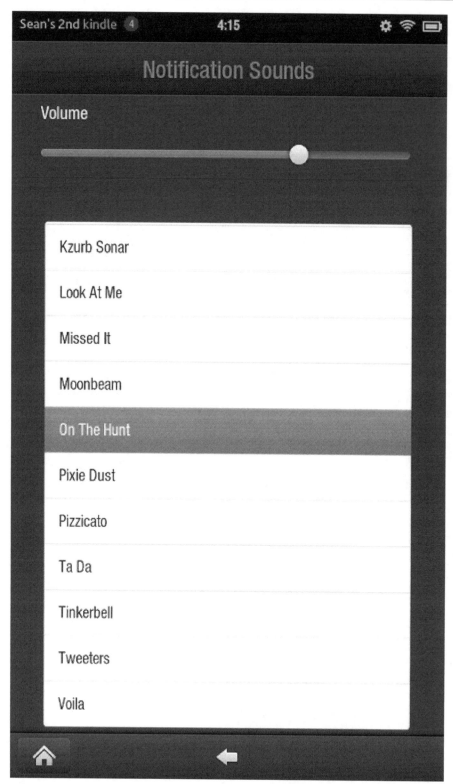

*Figure 6: Notification Sound Menu*

# 5. Turning Keyboard Sounds On or Off

The Kindle Fire can make a sound every time a key is touched. To turn keyboard sounds on or off:

1. Touch the ⚙ icon at the top right of the screen. The Quick Settings Banner appears.
2. Touch the ⊕ icon. The Settings screen appears.
3. Touch **Kindle Keyboard**. The Keyboard Settings screen appears, as shown in **Figure 7**.
4. Touch the `ON OFF` switch next to 'Sound on keypress'. The `ON OFF` switch appears and keyboard sounds are turned on.
5. Touch the `ON OFF` switch next to 'Sound on keypress'. The `ON OFF` switch appears and keyboard sounds are turned off.

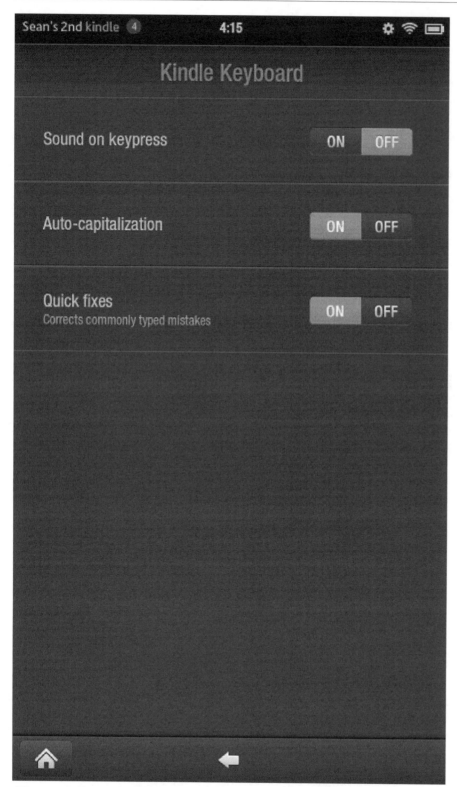

*Figure 7: Keyboard Settings Screen*

# 6. Turning the Lock Screen Password On or Off

Turning on a Lock Screen password prevents unauthorized users from accessing your Kindle Fire. To turn the Lock Screen Password on or off:

1. Touch the ⚙ icon at the top right of the screen. The Quick Settings Banner appears.
2. Touch the ⊕ icon. The Settings screen appears.
3. Touch **Security**. The Security Settings screen appears, as shown in **Figure 8**.
4. Touch the `ON OFF` switch next to 'Lock Screen Password'. The Lock Password screen appears, as shown in **Figure 9**.
5. Type the preferred password and touch the `OK` button. The Confirm Password text field is highlighted.
6. Type the same password again and touch the `OK` button. The Lock Screen password is set and the Lock Screen Password Prompt will appear every time you try to unlock the Kindle Fire, as shown in **Figure 10**.
7. Touch the `ON OFF` switch next to **Lock Screen Password**. The Password Prompt appears.
8. Enter the current password and touch the **Finish** button. The Lock Screen password is turned off.

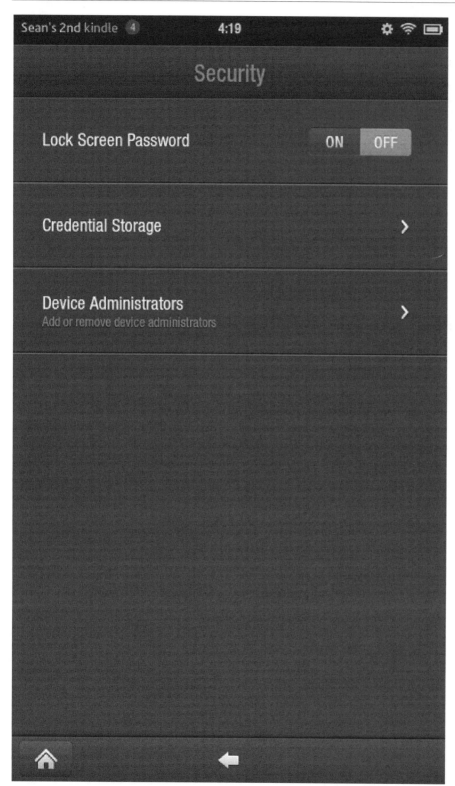

*Figure 8: Security Settings Screen*

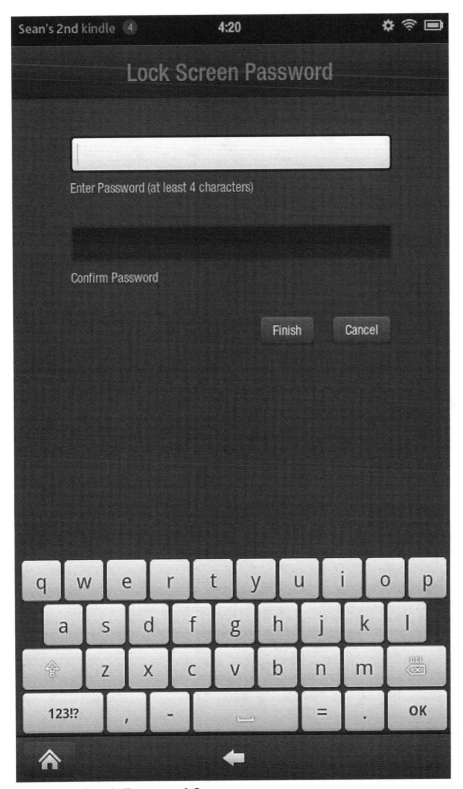

*Figure 9: Lock Password Screen*

*Figure 10: Lock Screen Password Prompt*

# 7. Setting the Date and Time Manually

The Kindle Fire will always automatically set the correct time and date after it is initially set up. To manually set the time and date:

1. Touch the ⚙ icon at the top right of the screen. The Quick Settings Banner appears.
2. Touch the ⊕ icon. The Settings screen appears.
3. Touch **Date & Time**. The Date & Time screen appears, as shown in **Figure 11**.
4. Touch the ⬛ON OFF switch next to 'Automatic'. Automatic Date & Time is turned off.
5. Touch **Set Time** and **Set Date** and use the ＋ and − buttons to set the correct time and date. The time and date are set.

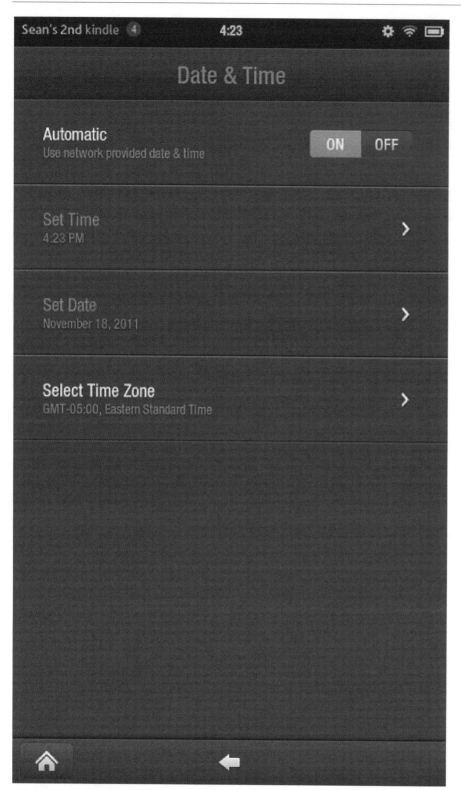

*Figure 11: Date & Time Screen*

# 8. Selecting the Time Zone

Selecting the time zone is the easiest way to set the clock if you are traveling or have moved. To select the Time Zone:

1. Touch the ⚙ icon at the top right of the screen. The Quick Settings Banner appears.
2. Touch the ⊕ icon. The Settings screen appears.
3. Touch **Date & Time**. The Date & Time screen appears.
4. Touch **Select Time Zone**. A list of time zones appears, as shown in **Figure 12**.
5. Touch a time zone. The new time zone is selected and the time is automatically updated.

*Figure 12: List of Time Zones*

# 9. Turning Auto-Capitalization On or Off

The Kindle Fire can automatically capitalize the first word of every sentence, names, and other commonly capitalized words. By default, Auto-Capitalization is turned on. To turn Auto-Capitalization on or off:

1. Touch the ⚙ icon at the top right of the screen. The Quick Settings Banner appears.
2. Touch the ⊕ icon. The Settings screen appears.
3. Touch **Kindle Keyboard**. The Keyboard Settings screen appears.
4. Touch the `ON OFF` switch next to 'Auto-capitalization'. The `ON OFF` switch appears and Auto Capitalization is turned off.
5. Touch the `ON OFF` switch next to 'Auto-capitalization'. The `ON OFF` switch appears and Auto Capitalization is turned on.

# 10. Turning Quick Fixes On or Off

Quick Fixes is a feature that automatically corrects common typos while you type. By default, Quick Fixes is turned on. To turn Quick Fixes on or off:

1. Touch the ⚙ icon at the top right of the screen. The Quick Settings Banner appears.
2. Touch the ⊕ icon. The Settings screen appears.
3. Touch **Kindle Keyboard**. The Keyboard Settings screen appears.
4. Touch the `ON OFF` switch next to 'Quick fixes'. The `ON OFF` switch appears and the Quick Fixes feature is turned off.
5. Touch the `ON OFF` switch next to 'Quick fixes'. The `ON OFF` switch appears and the Quick Fixes feature is turned on.

# 11. Turning Wi-Fi On or Off

The Kindle Fire can connect to a Wi-Fi network to access the internet and use such features as the Amazon store. However, Wi-Fi can also quickly drain the battery, so it should only be turned on when necessary. To turn Wi-Fi on or off:

1. Touch the ⚙ icon at the top right of the screen. The Quick Settings Banner appears.
2. Touch the 📶 icon. 'Wireless Networking' appears.
3. Touch the **ON OFF** switch next to 'Wireless Networking'. Wi-Fi is turned on and a list of networks appears, as shown in **Figure 13**. If you have set up Wi-Fi in the past, the Kindle Fire automatically connects to the saved network. Otherwise, proceed to step 4.
4. Touch a Wi-Fi network in the list. The password prompt appears.
5. Type the network password, which is usually found on the side or back of your wireless router. Touch the **Connect** button. Provided you have entered the correct password, the Kindle Fire connects to the Wi-Fi network.
6. Touch the **ON OFF** switch next to 'Wireless Networking'. Wi-Fi is turned off.

*Figure 13: List of Networks*

# Tips and Tricks

## Table of Contents

## 1. Maximizing Battery Life

There are several things you can do to increase the battery life of the Kindle Fire:

- Lock the Kindle Fire whenever it is not in use. To lock the device, press the **Power** button at the bottom of the device once.

- Keep the screen timeout feature set to a small amount of time to dim and turn off the screen when the Kindle Fire is idle. To learn how to change the screen timeout, refer to *"Setting the Screen Timeout"* on page 164.

- Turn down the brightness. To learn how to change brightness settings, refer to *"Adjusting the Brightness"* on page 162.

- Turn off Wi-Fi when not in use. To learn how to turn Wi-Fi off, refer to *"Turning Wi-Fi On or Off"* on page 182.

## 2. Checking the Amount of Available Memory

To check the amount of available memory at any time:

1. Touch the ▓ icon at the top right of the screen. The Quick Settings Banner appears.
2. Touch the ⊕ icon. The Settings screen appears.
3. Touch **Device**. The Device settings screen appears and the amount of available memory is shown at the top of the list.

## 3. Freeing Up Memory

There are several actions that can free up memory on the Kindle Fire. Try one or more of the following:

- Remove applications that are no longer needed from the device. Refer to *"Removing an Application from the Kindle Fire"* on page 159 to learn how.
- Remove all temporary internet files. To delete these files:

  1. Touch **Web** at the top right of the Library. The Silk browser opens.
  2. Touch the ▤ button at the bottom of the screen. The Browser menu appears.
  3. Touch the ✖ icon. The Browser Settings screen appears.
  4. Touch **Clear all cookie data**, **Clear cache**, or **Clear history**. The corresponding files are deleted.

## 4. Viewing the Full Horizontal Keyboard

The full horizontal keyboard provides much better accuracy than the vertical keyboard when typing. Simply rotate the Kindle Fire horizontally while entering text to turn on the horizontal keyboard, provided that screen orientation is unlocked.

## 5. Searching an eBook for a Word or Phrase Quickly

You can search for a word or phrase in an eBook without typing it. Touch and hold the word in the eBook. The Word menu appears. Touch **More** and then touch **Search in Book**. A list of locations where the word appears is shown. You can also search for a whole phrase by touching and holding a word and dragging your finger to select the rest of the phrase.

## 6. Viewing the Trailer for a Movie

In order to make a more informed decision when purchasing or renting a movie, you can view its cinematic trailer. To view a movie trailer, touch the Watch Trailer button in the movie description.

## 7. Setting a Download Default for Music

You can set music to automatically download either to your device or to your Amazon Cloud, therefore avoiding having to choose every time you purchase music. To set a download default for music:

1. Touch **Music** at the top of the Library. The Music library appears.
2. Touch the button at the bottom of the screen. The Music menu appears.
3. Touch the icon. The Music Settings screen appears.
4. Touch **Delivery preference**. The Delivery Preference menu appears.
5. Touch the preferred delivery method. The new delivery method is set.

## 8. Controlling Music from the Lock Screen

You can control music without having to unlock the Kindle Fire. To turn on Lock-Screen Controls:

1. Touch **Music** at the top of the Library. The Music library appears.
2. Touch the button at the bottom of the screen. The Music menu appears.
3. Touch the icon. The Music Settings screen appears.
4. Touch **Delivery preference**. The Delivery Preference menu appears.
5. Touch **Lock-screen controls**. Lock-Screen Controls are turned on and you can now control music without unlocking the device.

## 9. Replying to an Email Quickly

There is no need to open an email in order to reply to it. Instead, touch and hold the email in your Inbox. The Email menu appears. Touch **Reply**. A new email is created in reply to the original.

## 10. Marking an Email as Spam

In order to avoid receiving email from the same sender in the future, mark it as spam. To mark an email as spam, touch and hold the email in your Inbox. The Email menu appears. Touch **Mark as spam**. The email and all future emails from the same sender are moved to the Spam folder.

## 11. Viewing All Email from a Specific Sender

Instead of manually searching for all email sent by a single person, you can touch and hold an email in your Inbox and then touch **More from this sender**. All emails in your Inbox that were sent by the same person appear.

## 12. Closing All Tabs at Once in the Silk Browser

In addition to closing one tab at a time, you can also close all tabs simultaneously in the Silk browser. To close all tabs, touch and hold a tab and then touch **Close all tabs**. Once all tabs are closed, the Bookmarks screen appears.

## 13. Deleting a Bookmark in the Silk Browser

*Warning: Once a bookmark is deleted, it is gone for good. There is no confirmation dialog after you touch* **Delete***.*

Clean up your Bookmarks by deleting unneeded ones. To delete a bookmark from the Bookmarks screen, touch and hold it and then touch **Delete**. The Bookmark is deleted.

# 14. Closing Applications Running in the Background

When you return to the Library after using an application, it is left running in the background. Some applications take up a lot of memory and may slow down your Kindle Fire. To close an application running in the background:

1. Touch the ⚙ icon at the top right of the screen. The Quick Settings Banner appears.
2. Touch the ⊕ icon. The Settings screen appears.
3. Touch **Applications**. A list of running applications appears.
4. Touch an application. The Application Usage data appears.
5. Touch the Force Stop button. A confirmation dialog appears.
6. Touch the OK button. The application is closed.

*Note: Force-stopping an application will not damage it.*

# 15. Viewing the Back Issues of a Periodical

By default, only the latest issue of a periodical is shown in your Newsstand library. To view all issues of a periodical that you have received on your device, touch and hold the cover of a periodical and touch **Show Back Issues**. All previous issues appear. Touch and hold a cover of any issue and touch **Hide Back Issues**. Only the most recent issue is shown.

# Troubleshooting

## Table of Contents

## 1. Kindle Fire does not turn on

Try one of the following:

- **Recharge the battery** - Use the included wall charger to charge the battery. If the battery power is extremely low, the screen will not turn on for several minutes. Do NOT use the USB port on your computer to charge the Kindle Fire.

- **Replace the battery** - If you purchased the Kindle Fire a long time ago and have charged and discharged the battery approximately 300-400 times, you may need to replace the battery. You will need to contact Amazon to do so. Refer to *"What to do if your problem is not listed here"* on page 192 to learn how.

## 2. Kindle Fire is not responding

If the Kindle Fire is frozen or is not responding, try one or more of the following. These steps solve most problems on the Kindle Fire:

- **Restart the Kindle Fire** - Press and hold the Power button for 20 seconds and then release it.

- **Remove Media** - Some downloaded applications or music may freeze up the Kindle Fire. Try deleting some of the media after restarting the device. To learn how to delete an application, refer to *"Removing an Application from the Kindle Fire"* on page 159. You may also reset and erase all data at once by doing the following:

*Warning: Any erased data is not recoverable.*

1. Touch the ⚙ icon at the top right of the screen. The Quick Settings Banner appears.
2. Touch the ⊕ icon. The Settings screen appears.
3. Touch **Device**. The Device Settings screen appears.
4. Touch **Reset to Factory Defaults**. A confirmation dialog appears.
5. Touch the [Erase everything] button. All data is erased and the Kindle Fire is reset to factory defaults.

## 3. Kindle Fire battery dies too quickly

According to Amazon, the Kindle Fire provides up to eight hours of continuous reading or 7.5 hours of video playback. If you find that the battery is dying considerably faster, turn off Wi-Fi before plugging in the Kindle Fire to charge. Refer to *"Turning Wi-Fi On or Off"* on page 182 to learn how. Also, turn off Wi-Fi before locking the device and when you are not using it.

## 4. eBook has missing pages or poor formatting

Report the poor quality to Amazon by opening the eBook description in the store and touching **Would you like to report poor quality or formatting in this book** at the bottom of the screen.

# 5. Cannot access the Web although connected to a Wi-Fi Network

Some Wi-Fi networks, such as those in airports, coffee shops, or hotels, may not require a network password when connecting, but do require authentication once you open the Silk browser. If authentication is required, the ▦ icon appears at the top right corner of the screen. Touch **Web** and then enter the authentication password to connect to the internet.

# 6. Screen does not rotate

If the screen does not turn or the full horizontal keyboard is not showing when rotating the Kindle Fire, it may be one of these issues:

- The application does not support the horizontal view.
- The Kindle Fire is lying flat while being rotated. Hold the Kindle Fire upright for the view to change in applications that support it.
- Screen rotation is locked. Touch the ▦ icon at the top right of the screen and then touch the ▦ icon. The ▦ icon appears and screen rotation is unlocked.

# 7. Touchscreen does not respond as expected

If the touchscreen does not perform the desired functions or does not work at all, try the following:

- Remove the screen protector, if you use one.
- Make sure your hands are clean and dry. Oily fingers can make the screen dirty and unresponsive.
- Restart the Kindle Fire.
- Make sure the touchscreen does not come in contact with anything but skin. Scratches on the screen are permanent and may cause malfunction.
- Apply slightly more force when touching the screen, though not so much as to break it. The touchscreen on the Kindle Fire is not as responsive as most.

## 8. Computer does not recognize device

Use only the USB cable that came with your Kindle Fire to connect the device to your computer. Connect the device directly to the computer, since some USB hubs will not be able to recognize it.

## 9. Photo or video does not display

If the Kindle Fire cannot open a photo or video, the file type is most likely not supported. Supported file types include:

*Images*
- BMP
- GIF
- JPEG
- PNG

*Videos*
- OGG
- MP4
- VP8

## 10. What to do if your problem is not listed here

If you could not resolve your issue, contact customer service using one of the following methods:

If you are in the U.S., call **1-866-321-8851**.
If you are outside the U.S., call **1-206-266-0927**.
Email Kindle at **kindle-cs-support@amazon.com**.
Visit **http://www.amazon.com/kindlesupport**.

# Kindle Fire Survival Guide

# from MobileReference

**Author: Toly K**
**Editor: Julia Rouhart**

**This book is also available in electronic format from the following vendors:**

- **Amazon.com**
- **Barnesandnoble.com**
- **iBooks for iPhone and iPad**

MobileReference is a brand of SoundTells, LLC.
Please email questions and comments to support@soundtells.com. Normally we are able to respond to your email on the same business day.

MobileReference®. Intelligence in Your Pocket™.

## *Other Books from the Author of the Survival Guide Series, Toly K*

*iPad*

*HTC Incredible*

*Samsung Droid Fascinate*

*iPad 2*

*Nook Color*

*Sony Reader*

*Nook Simple Touch*

*How to Find & Download Free eBooks*

*iPad Apps for Scientists*

*Galaxy Tab*

*Droid X*

*Kobo*

HTC Droid 4G

iPad Games for Kids

iPhone 3G and 3GS

Nook

Xoom

Kindle

PC to Mac

Kobo Touch

Kindle 2011

iPhone 4

Made in the USA
Charleston, SC
09 March 2012

For Ashley x

Visit the author's website: www.ericjames.co.uk

Written by Eric James
Illustrated by Sara Sanchez
Designed by Nicola Moore

Published by Sourcebooks Jabberwocky, an imprint of Sourcebooks, Inc.
P.O. Box 4410, Naperville, Illinois 60567-4410
(630) 961-3900
Fax: (630) 961-2168
jabberwockykids.com

Date of Production: October 2017
Run Number: HTW_PO250717
Printed and bound in China (IMG)
10 9 8 7 6 5 4 3 2 1

# Tiny the
# New England
## Easter Bunny

Written by
Eric James

Illustrated by
Sara Sanchez

sourcebooks
jabberwocky

One bright Easter morning,
while out for a jog,

Tiny hears,

"HELP!

I AM STUCK
IN A LOG."

He scratches his head,
thinking, "Who could that be?
It sounded like Fluff!
I had better go see."

Fluff's in a log
with her feet in the air.
"Hey, Fluff, what on earth
are you doing in there?"

A sad little voice
from an echoey space
says, "I thought this would make
a good egg-hiding place."

"You poor Easter Bunny,"
says Tiny, while giggling.
"I'll get you back out,
just hold tight and stop wriggling!"

Tiny pulls hard,
using all of his might.
He tries and he tries,
but his friend is stuck tight.

"My eggs," sighs poor Fluff.
"Who'll deliver them now?"
"I'll do it," says Tiny.
Fluff laughs and asks,

# "How?!"

"Don't worry, dear Fluff.
Leave it all up to me.
I watched you last Easter.
How hard can it be?"

This bunny looks **funny**... Yes, something is **wrong!**

His feet are
**too**
# big
and his
nose is too
**long.**

His skin
isn't
**furry,**
it's wrinkled
and rough.

His tail is
**too thin,**
and it's
**NOT**
made of fluff.

He's traveled through **Salem**
and **Brideport** already.
He's all out of puff
and his legs feel unsteady.

He **hops**, then he stops, then he **hops** a bit more, then he stops all the **hopping**...

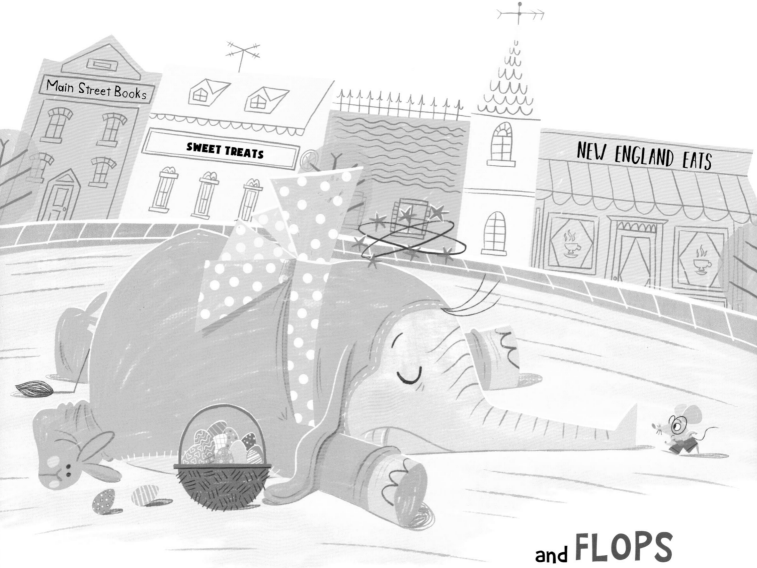

and **FLOPS** to the floor!

"Hello," squeaks a mouse
in his fake bunny ear.
"Oh my, how you've grown
since I met you last year.
I'm Marvin, remember?
You're running quite late...
I'll help if you like."
Tiny nods and says,

"Great!"

They head down to **Boston** and rush through the streets, delivering handfuls of chocolatey treats.

Next **Portland**, then **Providence**,
**Burlington** too.
There's not much time left
but there's SO MUCH to do!

"Speed up," Marvin squeaks,
"or we'll finish too late!
DIG under that hedge and
HOP over that gate."

This **Manchester** house
has a fence all around.
Poor Tiny tries digging
down into the ground.

But the hole is too small (or his body's too big).
"How odd," Marvin thinks. "I thought bunnies could dig!"

This large **Hartford** house
has a really high wall.
"Jump up," Marvin squeaks.
"C'mon, give it your all!"

So Tiny
jumps up but comes
down with a **THUMP!**
"How odd," Marvin thinks.
"I thought bunnies
could **JUMP!**"

"There's something not right,"
Marvin says. "Let me see..."
He scratches his chin and thinks,
"What can it be?"

"You're not very fast—
well, just look at those legs!
You're not very careful.
You've cracked half the eggs!"

"You do not have whiskers!
You're no good at hopping!
Those ears look quite fake,
and that's no bunny dropping!"

"Aha! Now I've got it!"
He jumps to his toes.
"No bunny is born with a
trunk for a nose!"

Tiny starts crying.
He wails, "Yes, it's true!
I thought I could do the
things real bunnies do."

"Don't worry. It's fine!"
Marvin squeaks, being nice.
"Do you mind if I offer
a little advice?"

"You need to start using
the talents you've got.
Be proud to be **you**.
Don't be something you're not!"

Boston Public Garden

"What talents?" says Tiny.
"What things can I do?"
He blows his big nose
and then aah...

aaah...

ACHOOOOOO!

"Eureka!" squeaks Marvin
from high in a tree.
"That wonderful,
big trunky nose
is the key!"

"It's strong and it's long.
It can pick things up too.
It's perfect for seeing
this Easter job through."

"We'll put it to work
just as soon as we can.
Let's head down to **New Haven**,
test out this plan."

This house has a fence,

and this house has a wall,

but with Tiny's big nose, there's no problem at all!

His long nose lifts up,
reaches over the top,
and he drops an egg down
on the lawn with a

P l O P!

And look at the hole
in this fence...that'll do!
There's room for an egg
and a trunk to fit through.

Now the job seems quite easy. (Well, that's how it goes
when an elephant uses his brains and his nose.)

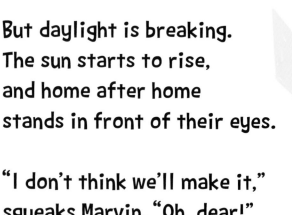

But daylight is breaking.
The sun starts to rise,
and home after home
stands in front of their eyes.

"I don't think we'll make it,"
squeaks Marvin. "Oh, dear!"
"Hang on," Tiny shouts.
"I've a marvelous idea!"

He sucks all the Easter eggs
into his nose,
and when his trunk's full
he takes aim...then he BLOWS!

Look at those eggs blasting out of his trunk, landing on lawns with a

THUNK!

THUNK!

THUNK!

THUNK!

The basket's soon empty.
"We did it, hooray!
Come on, let's help Fluff.
Oh, I hope she's okay."

"Hey, Fluff, Easter's saved.
I will get you out now!"

"Thank you,"
shouts Fluff.
"But you'll get me out...

how?"

At the side
of the pond,
Tiny dips in
his trunk.
He drinks and
he drinks
till the water's
all drunk!

And using
his nose
as a huge
water hose,
he blows
through the log...

Look at Fluff!

UP
she goes!

Happy Easter, New England!

How many Easter eggs
are in this picture?

This book belongs to:

_____

# Good Morning, Sweetie Pie

## And Other Poems for Little Children

CYNTHIA RYLANT

Illustrated by JANE DYER

Aladdin Paperbacks

New York   London   Toronto   Sydney

First Aladdin Paperbacks edition October 2004

Text copyright © 2001 by Cynthia Rylant
Illustrations copyright © 2001 by Jane Dyer

ALADDIN PAPERBACKS
An imprint of Simon & Schuster
Children's Publishing Division
1230 Avenue of the Americas
New York, NY 10020

Also available in a Simon & Schuster
Book For Young Readers hardcover edition.
Designed by Jane Dyer and Paul Zakris
The text of this book was set in 18-point Berkeley Oldstyle.
The illustrations are rendered in Winsor and Newton watercolors on
Waterford 140-pound hot press paper.

Manufactured in China
2  4  6  8  10  9  7  5  3  1

The Library of Congress has cataloged the hardcover edition as follows:
Rylant, Cynthia
Good morning, sweetie pie, and other poems for little children / by Cynthia
Rylant ; illustrated by Jane Dyer.
p. cm.
ISBN 0-689-82377-0 (hc)
1. Family—Juvenile poetry. 2. Children's poetry, America.
[1. Parent and child—Poetry. 2. Babies—Poetry. 3. American poetry.] I. Dyer,
Jane, ill. II. Title.
PS3568.Y55 G66 2002
811'.54—dc21
00-058785
ISBN 0-689-87060-4 (pbk.)

Jane Dyer would like to thank the people at
Smith College Child Care Center at Sunnyside for their assistance.

*To Cindy and her*
*beautiful grandbaby, Brooke*
*—C. R.*

*For my great nephew,*
*Andrew John Reimer,*
*the newest sweetie pie, with love*
*—J. D.*

# Contents

## Good Morning, Sweetie Pie

When the birds begin their singing
and the sun begins its sunning
and the morning glories
open up all blue . . .
there's a mama or a papa
or a gramma somewhere saying:
"Good morning, Sweetie Pie,
how are you?"

And a child is slowly waking,
slowly taking his sweet time,
he's been flying in his dreams
the whole night through.
But his little ears hear someone
and he knows it's someone dear
who is saying: "'Morning, Honey,
I love you."

There's a kiss upon a finger
and another on a nose
and a tickle on some tiny baby feet.
There's a wiggle under covers
and a giggle in the dark
and an "Oh, Cutie Dumpling, you're so sweet!"

And a child is peeking out now
on this warm and yellow day,
he is peeking at the dear one by his bed.
It is Mama or it's Papa,
maybe Gramma come to visit,
saying, "Where's that little Apple Sleepyhead?"

And the sleepyhead is up now,
he is rubbing sleepy eyes,
he is yawning big and stretching out his toes.
And his mama or his papa
or his gramma-come-to-visit
tells him, "Let's get us some crunchy Oaty-O's!"

So the child comes into morning
where the birds are gently singing
and the glories are still
blooming in the sun.
And he sits up at the table
with his dear one right beside him
who is saying:
"'Morning, Angel, you're my One."

And then Sweetie eats his breakfast
while the dear one calls him "Sugar"
and a puppy licks
his little baby toes.
It is morning in the kitchen
and there's nothing could be better
than a dear one
and some crunchy Oaty-O's!

## Baby Has a Sandbox

Baby has a sandbox,
he fills it up with trucks.
He adds a few long-neck giraffes
and little baby ducks.
He sprinkles in some little men
and little women, too.
Then stirs them with a little stick
to make a sandbox goo.
Baby plays there until ten
and then it's time to nap.
He carries everybody in—
poor Mama's sandy lap!
But Mama doesn't mind at all
her sandy baby's toys.
'Cause Mama knows that life is best
with sandy baby boys!

## Little Cutie-Face

Papa loves to give a ride
to little Cutie-Face.
He picks her up and gallops off
and takes her place to place.
He takes his Cutie up the hill
and gallops her back down.
He rides her to the castle
and around the castle-town.
He takes her to the ocean
on his bumpy papa-back,
then stands right in the middle
with his little Cutie-pack!
Papa and his Cutie-Face
go riding here and there,
and there is so much silliness
that neighbors stop and stare.

Then when the ride is over
Papa gallops in the house,
and sneaks two cookies and some milk
to his small Cutie-mouse.
Papa loves to give a ride,
it brightens up his day
to take his little Cutie-girl
upon his back and play.
And Cutie loves her papa so,
he's such a funny horse!
Of all the papas in the world,
she thinks he's best, of course!

## Baby Loves a Rainy Day

Baby loves a rainy day,
his mama keeps him in to play.
She brings out all his baby toys
(the special ones for baby boys).
He puts some people in a house
and reads a book about a mouse.
He rolls a ball across the room
and sweeps the carpet with a broom.
He stacks some toy blocks up high
and cuts a piece of dough-y pie.
Then Mama puts some music on
and Baby sings a baby-song.
Baby loves a rainy day
when Mama keeps him in to play.

## Going in the Car

Going in the car today
going into town,
little Girl and Papa
with the windows halfway down.
Waving to the paperboy,
waving to the train,
waving to the kitty cat,
waving to the plane.
Papa plays the radio
and sings a Papa song.
Little Girl and Mrs. Bear,
they always sing along.
Stopping at the grocery store,
stopping for the mail,
stopping at the hardware now,
'cause Papa needs a nail.

Going in the car today
going into town.
Little Girl and Papa
with the windows halfway down.

## Baby Has a Bath Today

Baby has a bath today,
he takes his teddy bear.
He fills his hands with baby soap
and washes Teddy's hair.
His mama washes Baby's ears
and Baby's pretty neck.
She rubs a cloth on Baby's legs
and gives his toes a check.
When Baby and his teddy
are all bright and squeaky clean,
Mama dries them off and then
they eat a tangerine.

## Sweetie's Messes

At suppertime when Sweetie eats
She surely makes a mess,
With carrots in her curly hair
And green beans on her dress.
Papa tries to keep her clean
But he's no help at all,
His hair is full of apples
And a meatball's down the hall.
The puppy loves when Sweetie eats,
He licks up her spaghetti.
If Sweetie drops a tasty noodle,
Puppy's always ready!
Suppertime with baby girl
Leaves Papa looking funny.
He's got some squash in his left ear
And one shoe's full of honey!
But Papa loves this baby so,
He doesn't mind her messes.
He'll just keep washing Sweetie's face
And washing Sweetie's dresses!

## Sleepy-boy

"Time for bed,"
says Mama-dear,
who picks up Sleepy-boy.
She finds his favorite blanket
and his favorite
sleeping toy.
She puts him
in his jammies
and she carries him to bed.
Then covers him
with little stars
and kisses his sweet head.
She turns his little
moon-light on
and lines up every bear.
She puts the horse
and donkey in
the stable that they share.
Now animals and boys
can sleep safe the whole night through.
'Cause Mama's watching over them . . .
that's what mamas do.

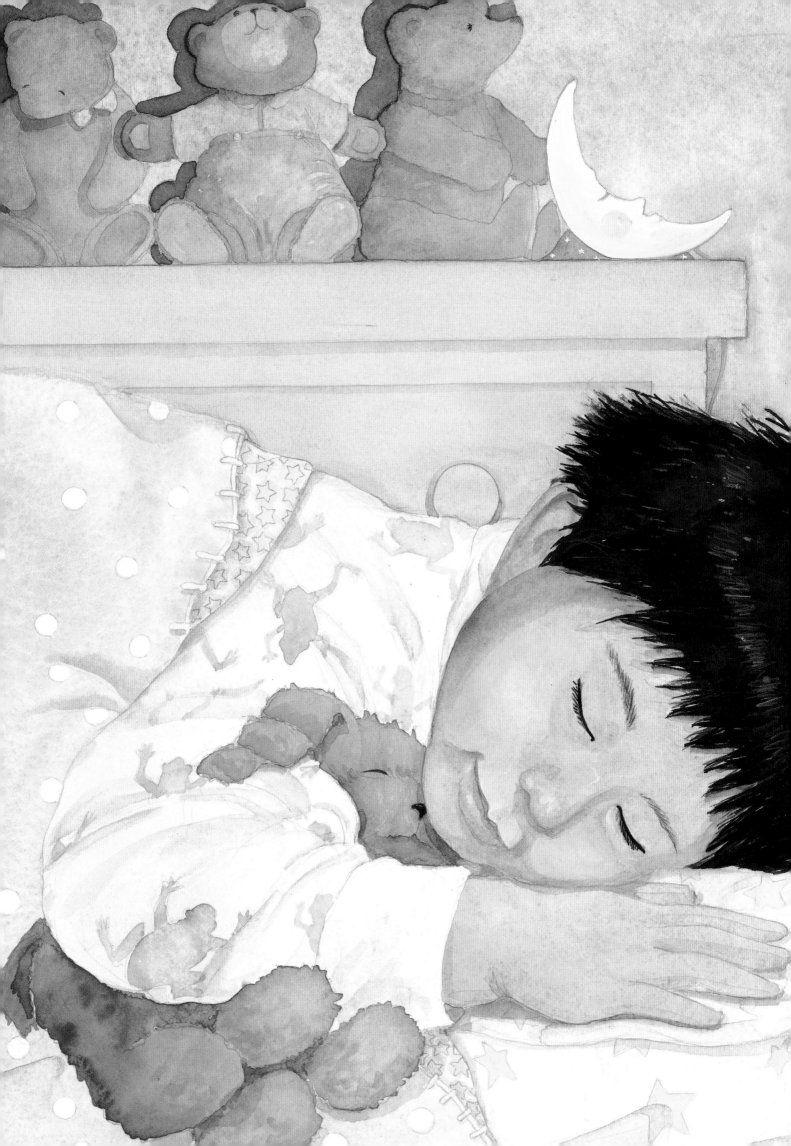

# Special Thanks to These Sweetie Pies

## MINH
*Contents*

## LEON
*Good Morning, Sweetie Pie*

## EZRA
*Baby Has a Sandbox*

## AURORA
*Little Cutie-Face*

## AMAL
*Baby Loves a Rainy Day*

## AISHA
*Going in the Car*

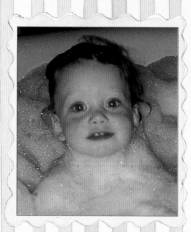

## WILLY
*Baby Has a Bath Today*

## ISABELLA
*Sweetie's Messes*

## ALEX
*Sleepy-boy*